HEINEMANN MATHEMATICS 6

Workbook

These are the different types of pages and symbols used in this book.

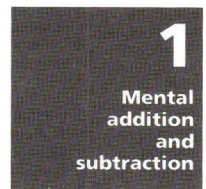

Workbook pages are of this type. They deal with mathematical concepts, skills and applications in number, measure, shape and handling data.

Problem solving

Some pages, or parts of a page, provide an opportunity for problem solving or investigative work.

Where a calculator would be useful this is indicated by a calculator symbol.

This symbol indicates that more work of this kind can be found on the numbered Reinforcement Sheet.

The work on this page is supported by the numbered Home Link-up activity.

Washed ashore

Sean and Dorothy find 34 tins of food.
Roy and Elaine find 28 tins.

34 add 28
34 + 20 = 54
54 + 8 = 62

Altogether we found **62 tins** of food.

1 Add **mentally**.

25 + 16 = ____   29 + 23 = ____   26 + 17 = ____   38 + 46 = ____
47 + 17 = ____   33 + 28 = ____   18 + 57 = ____   25 + 37 = ____
24 + 48 = ____   54 + 37 = ____   27 + 46 = ____   48 + 29 = ____
38 + 38 = ____   45 + 35 = ____   61 + 29 = ____   59 + 27 = ____

Sean finds a crate of orange juice.
He takes out 35 cartons which are empty.

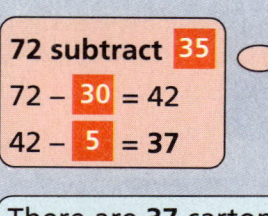

orange juice
72 cartons

72 subtract 35
72 − 30 = 42
42 − 5 = 37

There are **37** cartons left in the crate.

2 Subtract **mentally**.

42 − 27 = ____   66 − 38 = ____   75 − 46 = ____   66 − 57 = ____
53 − 25 = ____   81 − 34 = ____   34 − 15 = ____   77 − 29 = ____
47 − 28 = ____   78 − 49 = ____   91 − 66 = ____
93 − 77 = ____   60 − 32 = ____   84 − 48 = ____
82 − 53 = ____   96 − 76 = ____   80 − 17 = ____

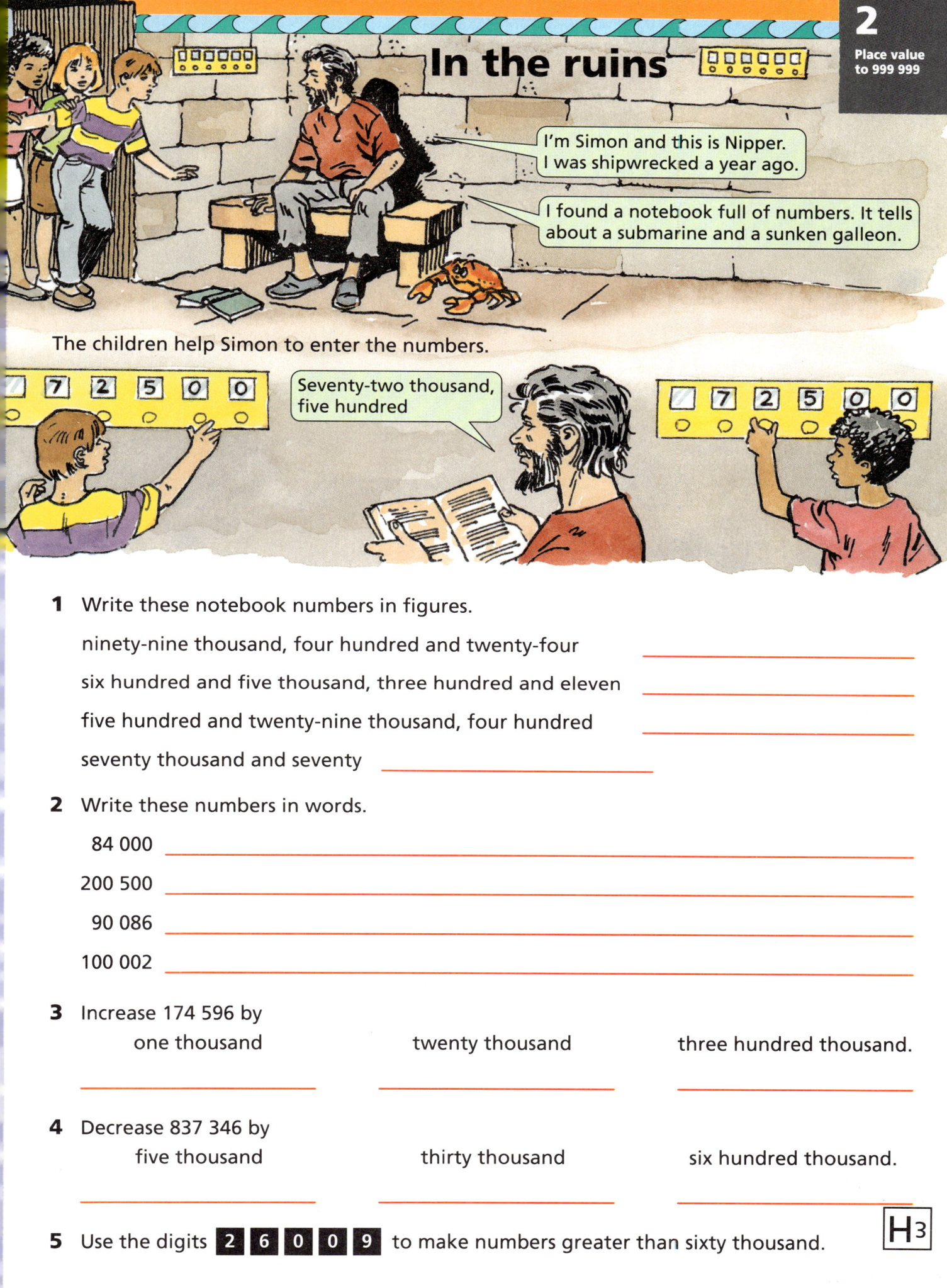

In the ruins

Place value to 999 999

I'm Simon and this is Nipper. I was shipwrecked a year ago.

I found a notebook full of numbers. It tells about a submarine and a sunken galleon.

The children help Simon to enter the numbers.

Seventy-two thousand, five hundred

1 Write these notebook numbers in figures.

ninety-nine thousand, four hundred and twenty-four _____

six hundred and five thousand, three hundred and eleven _____

five hundred and twenty-nine thousand, four hundred _____

seventy thousand and seventy _____

2 Write these numbers in words.

84 000 _____

200 500 _____

90 086 _____

100 002 _____

3 Increase 174 596 by

    one thousand   twenty thousand   three hundred thousand.

_____   _____   _____

4 Decrease 837 346 by

    five thousand   thirty thousand   six hundred thousand.

_____   _____   _____

5 Use the digits **2 6 0 0 9** to make numbers greater than sixty thousand.

_____   _____   _____

H3

3 Place value to 999 999

The door

1 Complete.

1 more than 9999	1 less than 25 000	10 more than 26 890
_____	_____	_____

10 less than 23 800	100 more than 29 900	100 less than 31 000
_____	_____	_____

Problem solving

2 Use the digits 2 2 4 6 8
Find the number less than 70 000 which has
- the same units and hundreds digits
- its thousands digit double its units digit.

☐☐☐☐☐

3 Find the first number greater than 25 387 where
(a) its thousands and units digits are the same

_____

(b) its thousands and tens digits are the same

_____

(c) all its digits are odd.

_____

4 (a) Arrange all the answers from questions 1, 2 and 3 in order. Start with the smallest.

(b) Cross out all answers which
- are greater than 30 000
- are less than 25 000
- divide exactly by 10.

(c) The remaining number opens the door. What is it?

Go to Textbook page 3.

Start up signs

1 Colour in squares to find these start up signs.

(a) Start at 3. Add 3 each time.

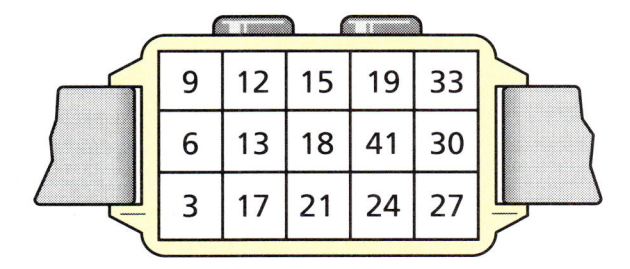

9	12	15	19	33
6	13	18	41	30
3	17	21	24	27

3, 6, 9, 12, ... are called **multiples** of 3.

(b) Start at 6. Add 6 each time.

18	24	30	39	54	60	66
12	17	36	42	48	65	72
6	9	14	21	43	52	78

Write the multiples of 6.

_____

2 (a) Colour multiples of
• 2 yellow • 3 green.

1	2	3	4	5
6	7	8	9	10
11	12	13	14	15
16	17	18	19	20
21	22	23	24	25
26	27	28	29	30

Communicator

(b) Colour multiples of • 2 red • 5 blue.

1	2	3	4	5	6
7	8	9	10	11	12
13	14	15	16	17	18
19	20	21	22	23	24
25	26	27	28	29	30
31	32	33	34	35	36
37	38	39	40	41	42
43	44	45	46	47	48
49	50	51	52	53	54

Communicator

(c) List the numbers coloured

• **yellow and** green _____

These are multiples of _____

• **red and** blue _____

These are multiples of _____

3 (a) Colour the multiples of
• 4 blue • 8 yellow.

(b) The numbers coloured blue **and** yellow are multiples of _____

1	2	3	4	5	6	7	8	9	10
11	12	13	14	15	16	17	18	19	20
21	22	23	24	25	26	27	28	29	30
31	32	33	34	35	36	37	38	39	40

4 Multiplication: multiples, table links

5 Hot lines

Multiplication: tables, digital roots

> The sum of the digits of **56** is 5 + 6 → 11.
> The sum of the digits of **11** is 1 + 1 → 2.   2 is the **digital root** of 56.

1 (a) Complete each table.

3 times table	
multiple	digital root
3	
6	
9	
12	
15	
18	
21	
24	
27	
30	

6 times table	
multiple	digital root
6	
12	
18	
24	
30	
36	
42	
48	
54	
60	

(b) In the red dial, the digital roots of the multiples of the 3 times table have been joined in order 3 → 6 → 9 → 3.

Use the blue dial. Join, in order, the digital roots of the multiples of the 6 times table.

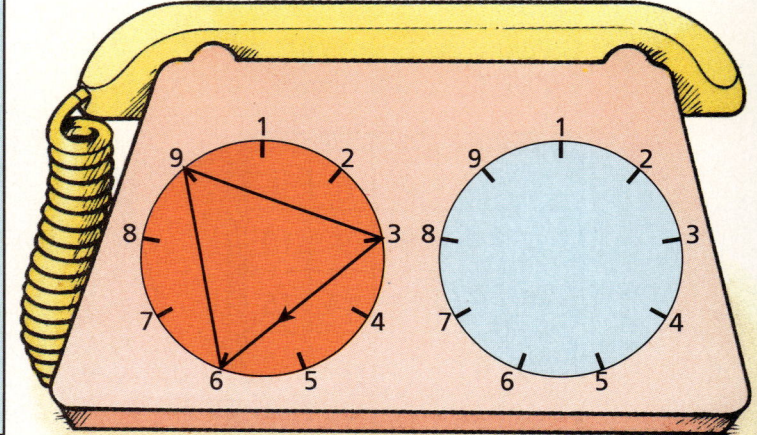

(c) What do you notice about the design on each dial? _____

2 (a) Complete.

2 times table	
multiple	digital root
2	
4	
6	
8	
10	
12	
14	
16	
18	
20	

7 times table	
multiple	digital root
7	
14	
21	
28	
35	
42	
49	
56	
63	
70	

(b) Join, in order, the digital roots of the multiples of the
- 2 times table
- 7 times table.

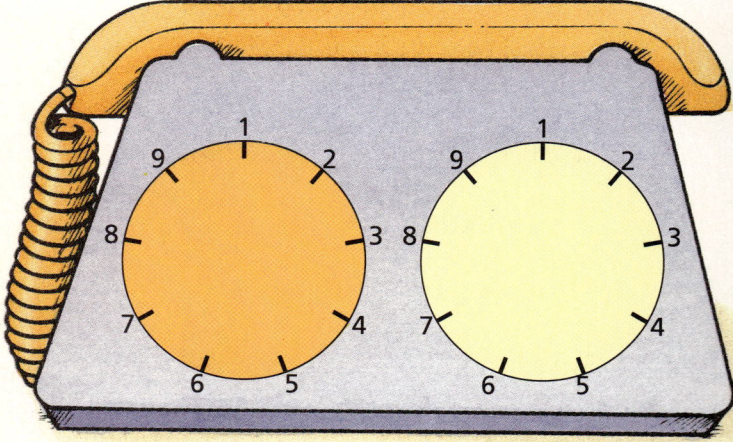

(c) What do you notice about the design on each dial? _____

Go to Textbook page 13.

Nicked!

1 Help Mog to complete her files about the robbers caught by Superhero.
  (a) Do each calculation.
  (b) The **largest** answer is the file number. Write it on the robber's photo.

9516 ÷ 3 _____	5785 ÷ 5 _____	6832 ÷ 4 _____	4072 ÷ 8 _____
9368 ÷ 8 _____	9863 ÷ 7 _____	9018 ÷ 9 _____	8406 ÷ 6 _____
5895 ÷ 9 _____	8869 ÷ 7 _____	4578 ÷ 6 _____	9576 ÷ 2 _____

Dicky Fingers  Martha Moll  Colin Conman  Deena Dip

2 (a) Complete the first three rows in this table.
  (b) Look for a pattern and complete the table.
  (c) Check your answers for row **R** to row **E**.
  (d) The robber's number is the answer in every row. Write this number on his photo.
  (e) Use the letter code to find the robber's name. Write his name on his photo.

C →	8	9	9	1	÷	9	=
N →	7	9	9	2	÷	8	=
H →	6	9	9	3	÷	7	=
R →							
S →							
A →							
T →							
E →							

4995	7992	3996	2997	8991	6993	1998	5994
↓	↓	↓	↓	↓	↓	↓	↓
___	___	___	___	___	___	___	___

Problem solving

3 The reward for catching Tee Leaf
  • is £9950 **to the nearest £10**
  • has both 7 **and** 9 as factors.
  (a) Find the **exact** amount of the reward.
  (b) Write this amount on his photo.

Reward £ _____

Tee Leaf

Go to Textbook page 26.

7 Division: rounding to the nearest unit

The aircraft's computer

Superhero and Mog do important calculations for the captain.

1 Complete each calculation, each sentence and each number line.

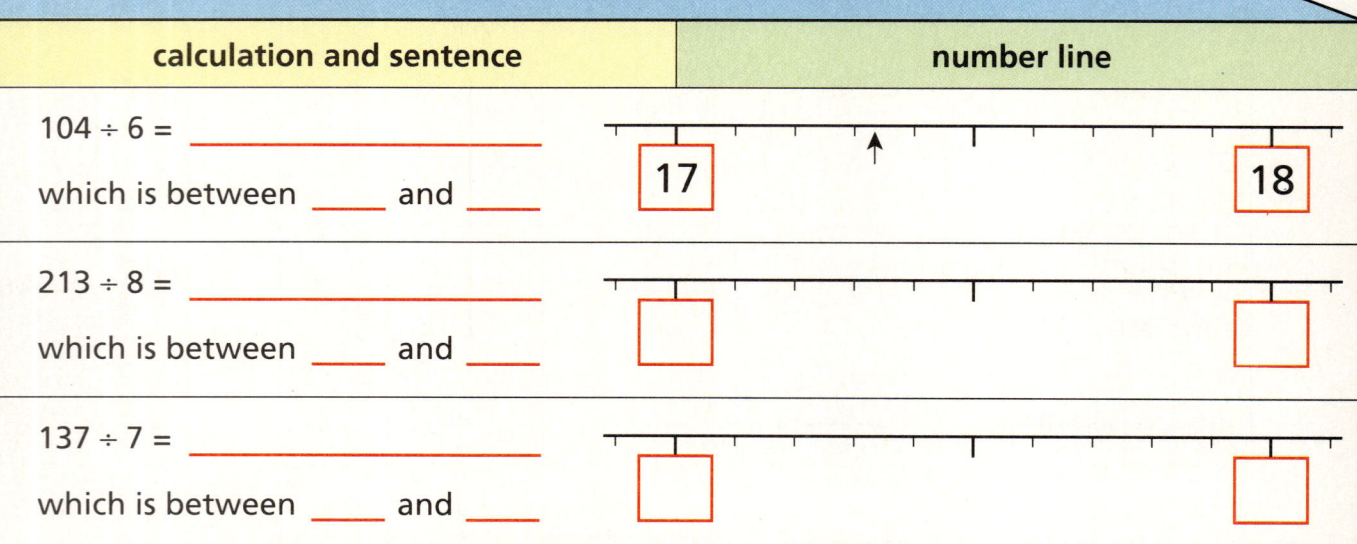

calculation and sentence	number line
104 ÷ 6 = _____ which is between ____ and ____	17 … 18
213 ÷ 8 = _____ which is between ____ and ____	□ … □
137 ÷ 7 = _____ which is between ____ and ____	□ … □

Check that 121 ÷ 8 = 15·125

15 — 16

121 ÷ 8 is between 15 and 16
nearer 15
15 to the nearest unit

Check that 202 ÷ 7 = 28·857142

28 — 29

202 ÷ 7 is between 28 and 29
nearer 29
29 to the nearest unit

2 Complete each calculation, each sentence and each number line.

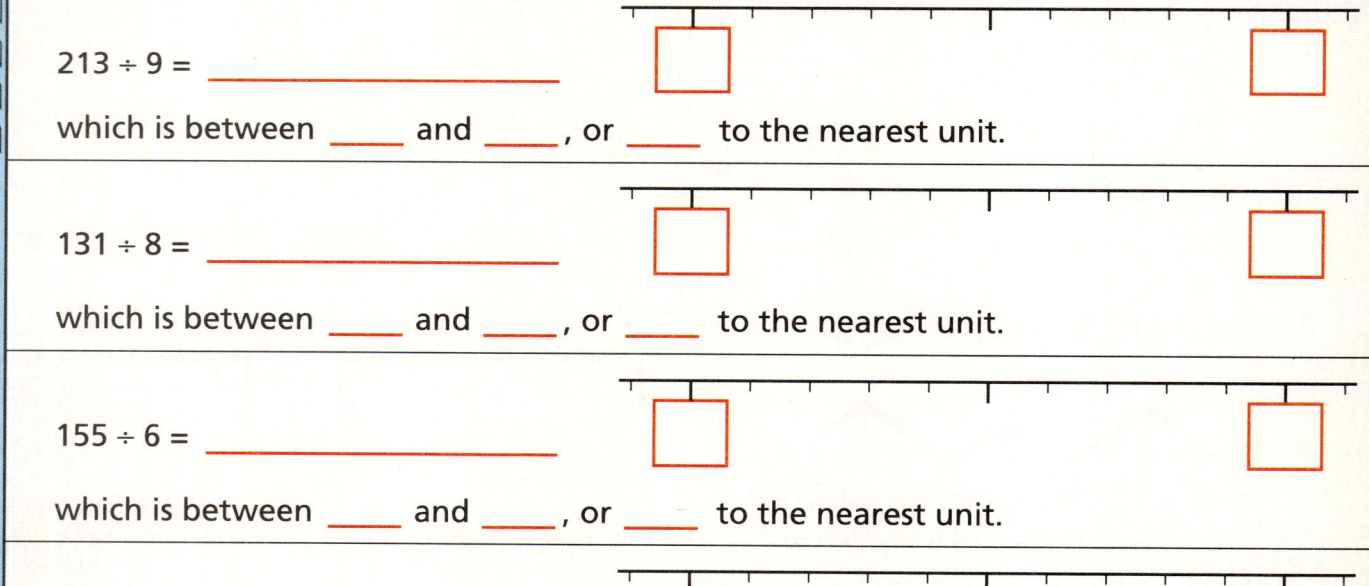

213 ÷ 9 = _____
which is between ____ and ____, or ____ to the nearest unit.

131 ÷ 8 = _____
which is between ____ and ____, or ____ to the nearest unit.

155 ÷ 6 = _____
which is between ____ and ____, or ____ to the nearest unit.

331 ÷ 7 = _____
which is between ____ and ____, or ____ to the nearest unit.

Go to Textbook page 27.

The Great Hall

8
Fractions: equivalence

1 Colour **one half** of each design. The first pair are done for you.

(a)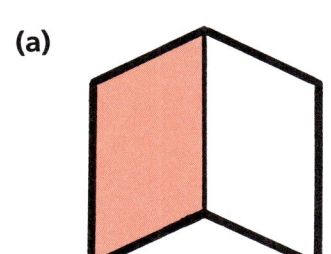

Complete: $\dfrac{1}{2} = \dfrac{}{4}$

(b)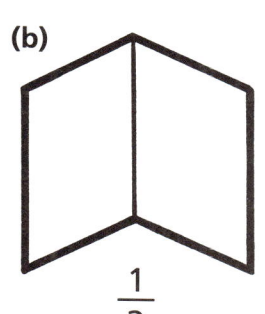

$\dfrac{1}{2} = \dfrac{}{6}$

(c)

$\dfrac{1}{2} = \dfrac{}{8}$

(d)

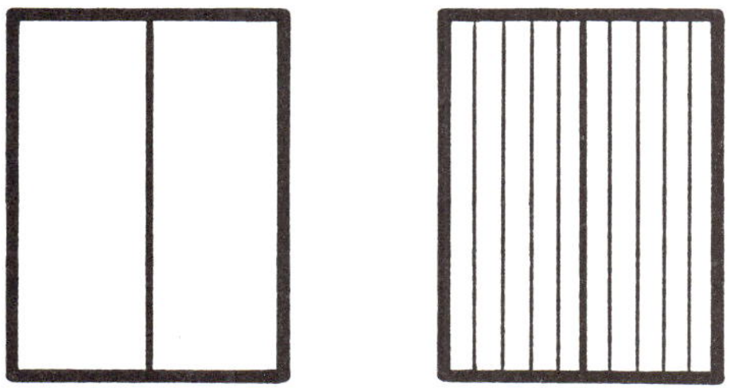

$\dfrac{1}{2} = \dfrac{}{10}$

2 Colour **one quarter** of each design. Complete the fractions.

(a)

$\dfrac{1}{4} = \dfrac{}{8}$

(b)

$\dfrac{1}{4} = \dfrac{}{20}$

3 Colour **one fifth** of each design. Complete the fractions.

(a)

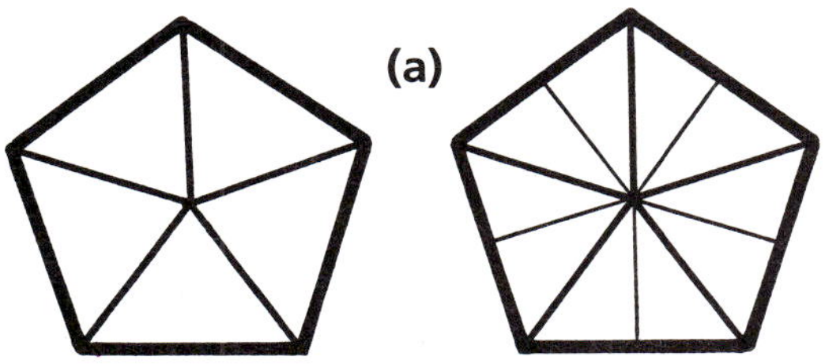

$\dfrac{1}{5} = \dfrac{}{10}$

(b)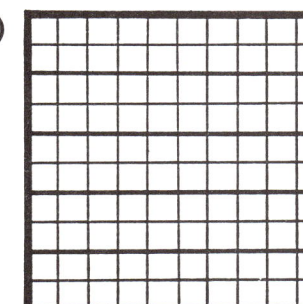

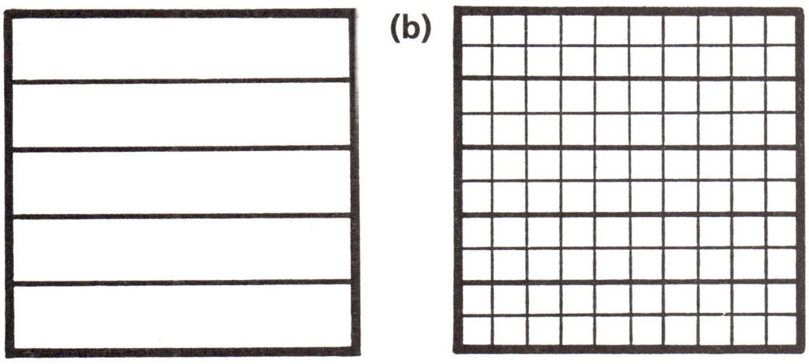

$\dfrac{1}{5} = \dfrac{}{100}$

Go to Textbook page 31.

9 Fractions: equivalence

The Great Hall again

1 Colour the second design of each pair to match the first one. Complete the fractions.

(a) 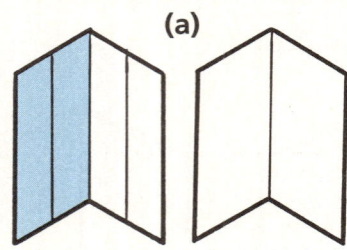 $\frac{2}{4} = \frac{}{2}$

(b) 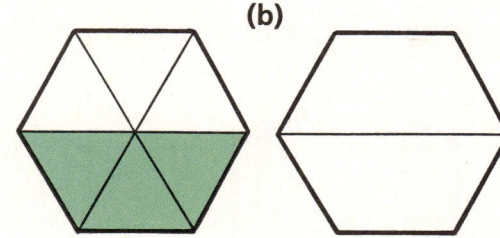 $\frac{3}{6} = \frac{}{}$

(c) 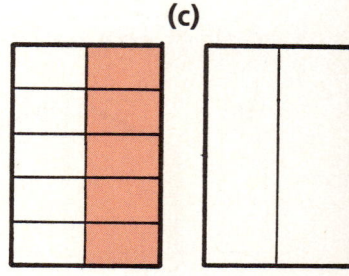 $\frac{5}{10} = \frac{}{}$

(d) 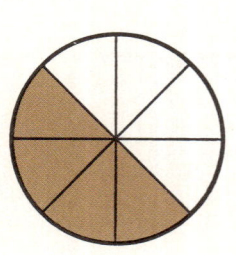 $\frac{4}{8} = \frac{}{}$

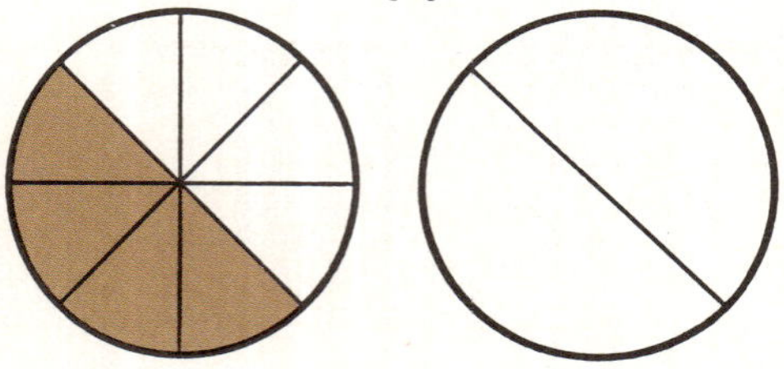

(e) 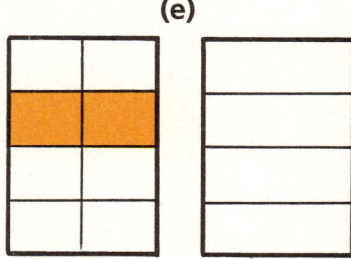 $\frac{2}{8} = \frac{}{}$

(f) 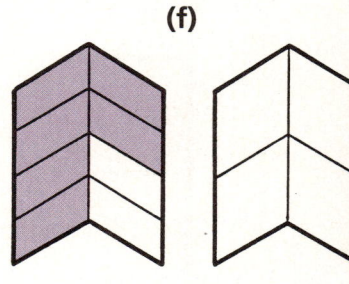 $\frac{}{8} = \frac{}{4}$

(g) 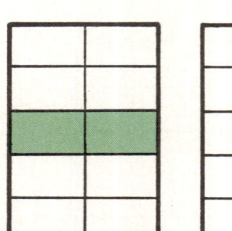 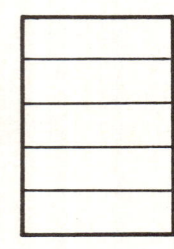 $\frac{2}{10} = \frac{}{5}$

(h) 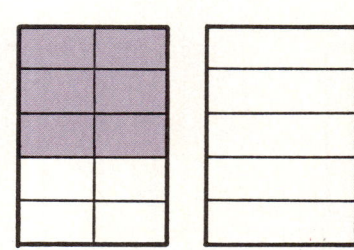 $\frac{6}{10} = \frac{}{}$

(i) 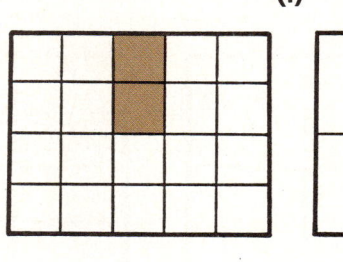 $\frac{}{20} = \frac{}{}$

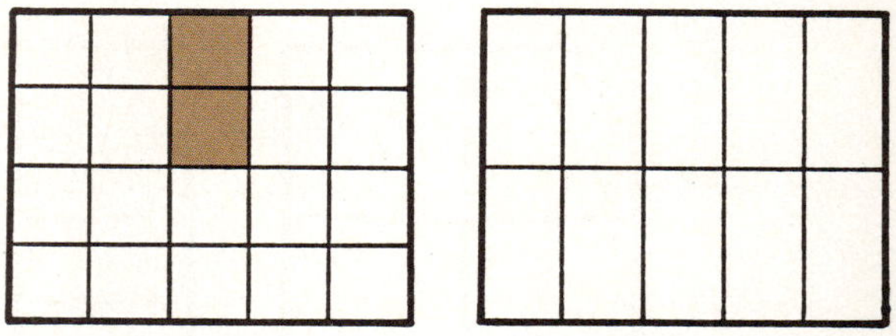

(j) 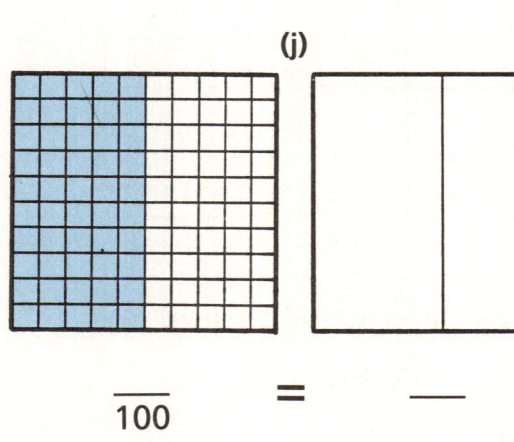 $\frac{}{100} = \frac{}{}$

(k) 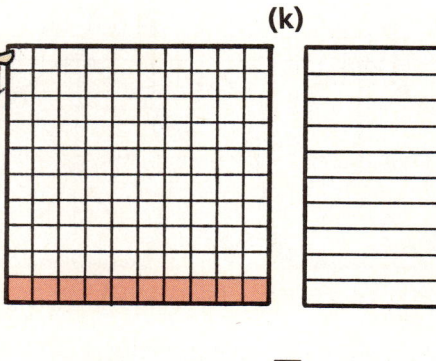 $\frac{}{100} = \frac{}{}$

Go to Textbook page 32.

Signs

10 Decimals: second decimal place, notation

1. Each sign has 100 light cells.
   Colour the given fraction of each sign.

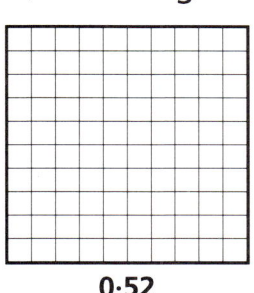 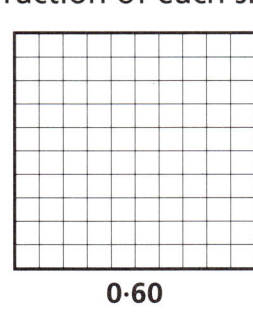

0·52   0·60

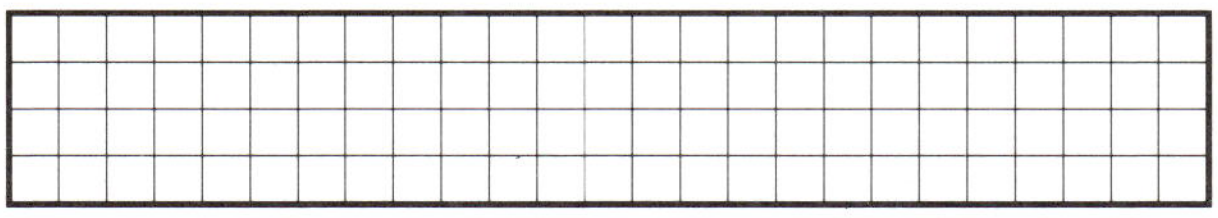

0·08

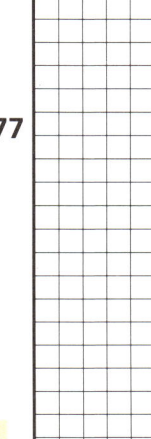

0·77

One whole sign and 34 hundredths of the other sign is coloured.

In decimal form
1 whole and 34 hundredths is 1·34

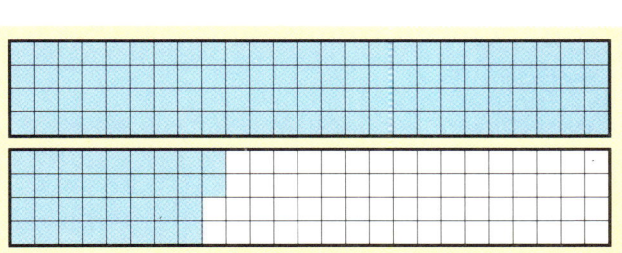

2. Write in decimal form the amount coloured.

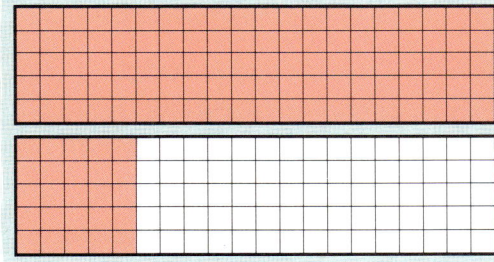   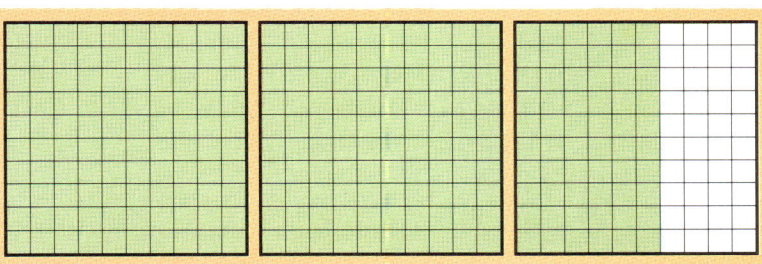

_____   _____

3. Colour to show the given amount.   2·06

1·61

   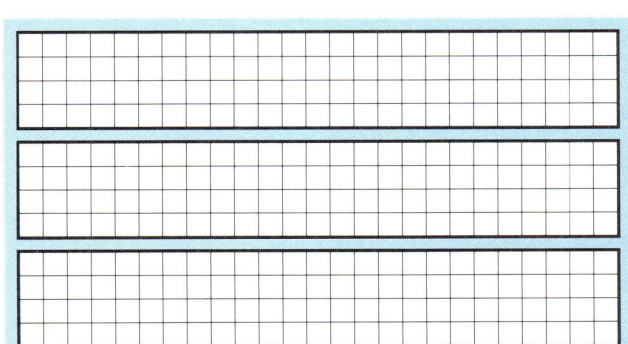

4. Write in decimal form:

   3 units and 14 hundredths _____   4 units and 99 hundredths _____

   5 units and 80 hundredths _____   3 units and 3 hundredths _____

Go to Textbook page 43.

R 12 H 27

11 Decimals: place value using a calculator

Shifting sands

1 Each year Alex measures the post-lengths sticking out of the sand.

Alex finds the change from year to year like this:

- He enters last year's length
- He decides what to add or subtract to give ...... this year's length.

`1.87`   `– 0 . 0 2` → `1.85`

Complete Alex's table for the other posts.

Last year Length in metres	Change in metres	This year Length in metres
1·87	– 0·02 →	1·85
1·43	→	1·13
2·94	→	2·99
2·51	→	1·51
1·36	→	1·76
0·56	→	1·56
0·18	→	0·12
1·45	→	1·4

2 (a) Alex's helpers each try to collect £10 for the Save the Sand Dunes fund. Jan has collected £4·26. Make £4·26 up to £10 by adding hundredths, **then** tenths, **then** units like this:

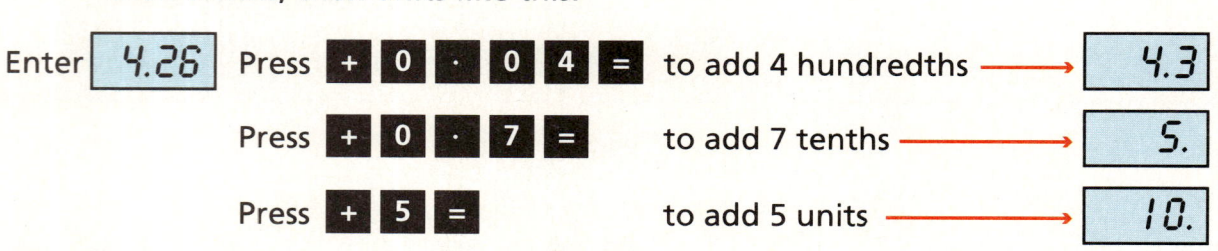

Enter `4.26`   Press `+ 0 . 0 4 =` to add 4 hundredths → `4.3`

Press `+ 0 . 7 =` to add 7 tenths → `5.`

Press `+ 5 =` to add 5 units → `10.`

The numbers **added** can be shown like this:

Enter	Add	Add	Add	Make
4.26 →	0·04 →	0·7 →	5 →	10.

(b) In the same way, make £10 starting with these numbers:

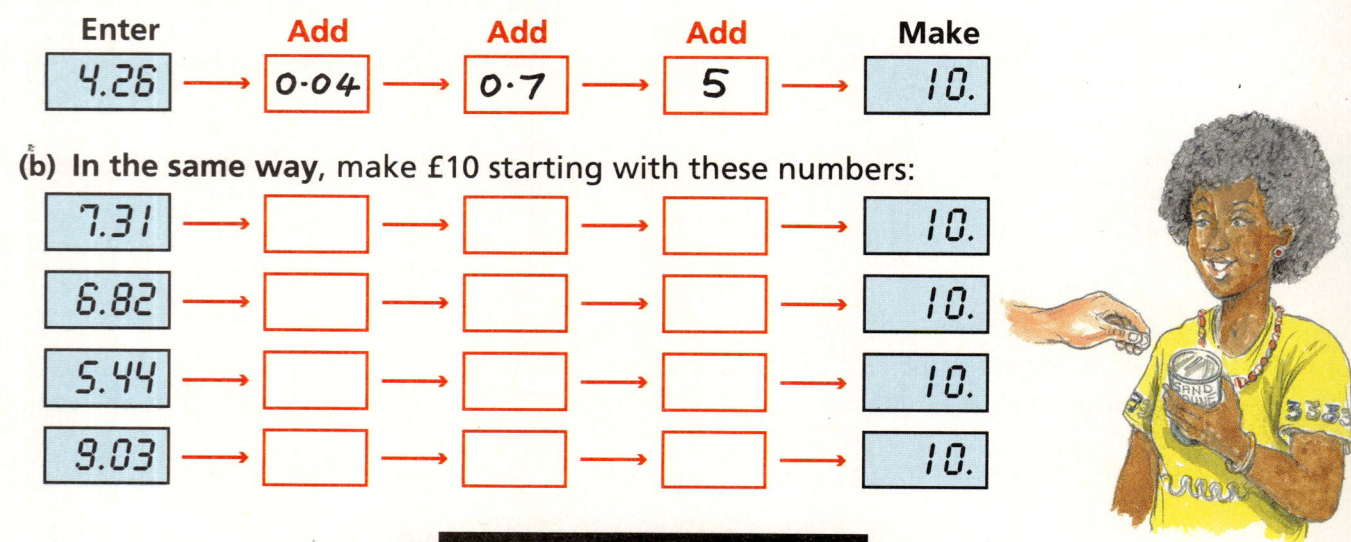

7.31 → ☐ → ☐ → ☐ → 10.

6.82 → ☐ → ☐ → ☐ → 10.

5.44 → ☐ → ☐ → ☐ → 10.

9.03 → ☐ → ☐ → ☐ → 10.

Go to Textbook page 45.

The Castle windows

12
Percentages: concept

Each window has 100 panes.

1 (a) Complete:

10% = $\frac{}{100}$ 50% = $\frac{}{100}$

(b) Colour 10% of the window red.
Colour 50% of the window blue.

(c) What percentage is **not** coloured? _____

2 Colour each window to show the given percentages.

red   30%
blue   25%
green   5%

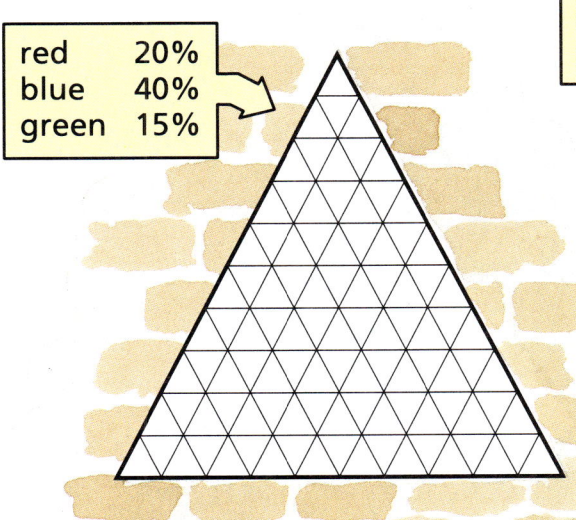

red   20%
blue   40%
green   15%

red   32%
blue   35%
green   18%

3 (a) Use red, blue and green to colour your own design for this window.

(b) Complete: _____ % is coloured red

_____ % is coloured blue

_____ % is coloured green.

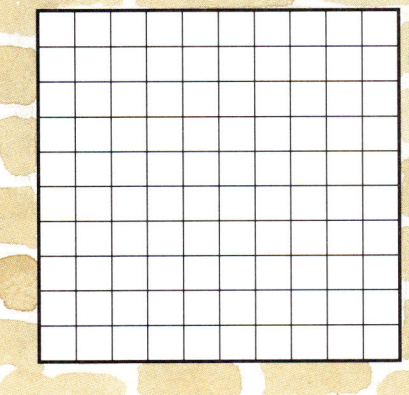

R 15   H 32

13 Percentages: 100% is one whole

The Castle Fund

$100\% = \frac{100}{100} = 1$ whole

100% of something is the whole of it.

The chart shows that, by 1st February, 25% of the money needed to restore the castle had been raised.

Castle Fund

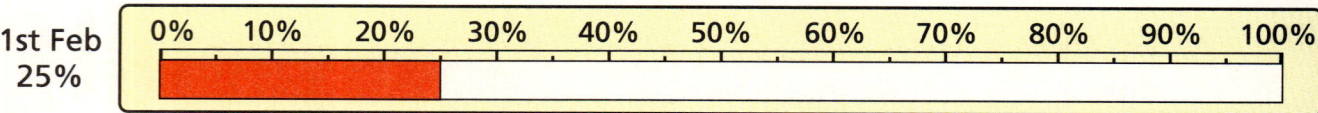

1st Feb 25%

1 (a) Colour each chart to show the percentage raised by the given date.

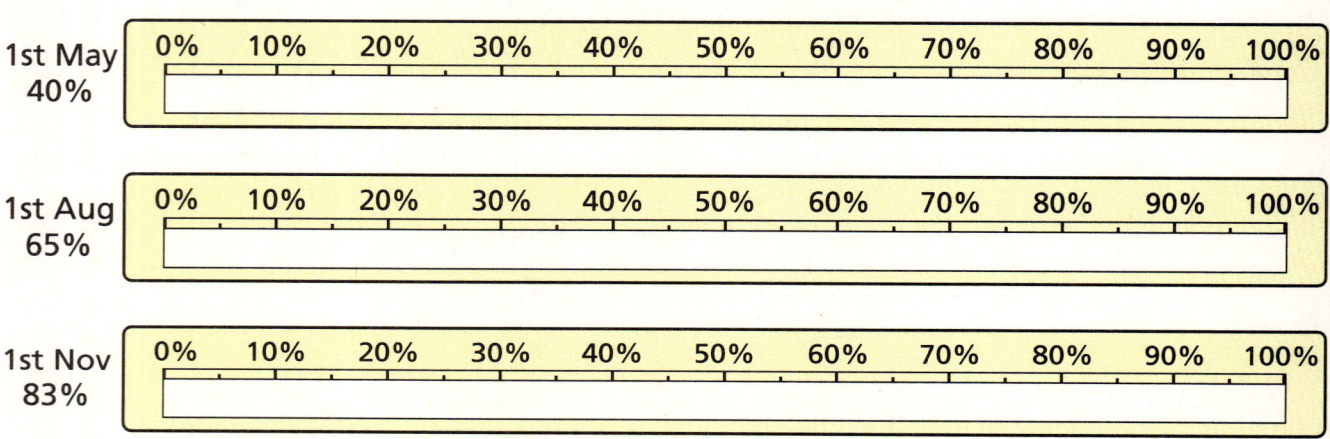

1st May 40%

1st Aug 65%

1st Nov 83%

(b) Complete the table to show the percentage still to be raised at each date.

	1st Feb	1st May	1st Aug	1st Nov
Percentage raised	25%			
Percentage to be raised				

2 Find the missing percentages.

Visitors' language	
English	73%
Not English	____ %

Types of visitor	
Senior citizens	12%
Other adults	54%
Children	____ %

Visitors' countries	
Britain	48%
Europe (rest of)	19%
America	20%
Japan	5%
Other countries	____ %

3 42% of visitors give money to the Fund. What percentage of the visitors do not give money? _____

The Royal Bedroom

There are 100 patches in each quilt.

1 (a) Colour $\frac{1}{2}$

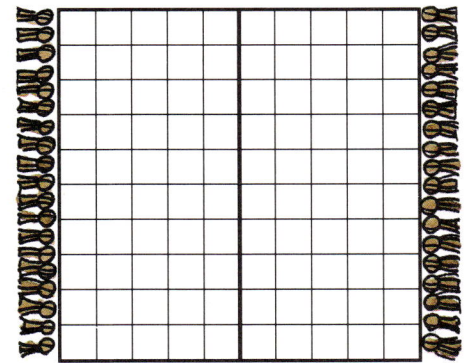

(b) Complete: $\frac{1}{2} = \frac{}{100} = \underline{}$ %

2 (a) Colour $\frac{1}{4}$

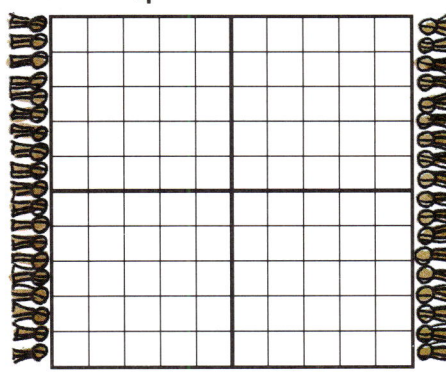

(b) Complete: $\frac{1}{4} = \frac{}{100} = \underline{}$ %

3 (a) Colour $\frac{1}{10}$

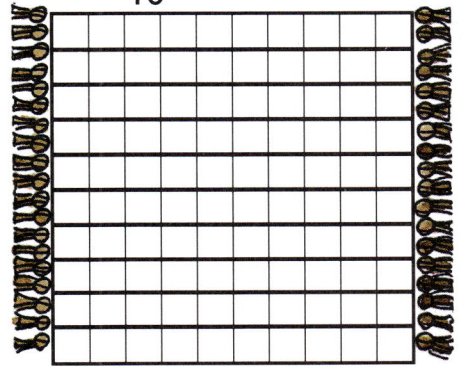

(b) Complete: $\frac{1}{10} = \frac{}{100} = \underline{}$ %

4 (a) Colour $\frac{3}{4}$

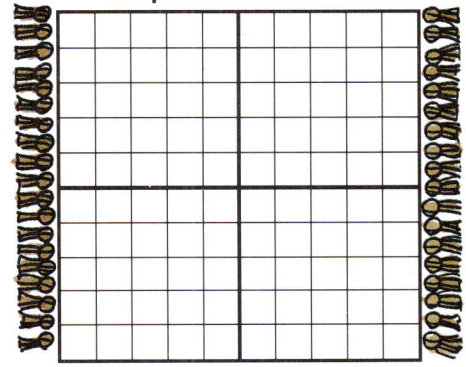

(b) Complete: $\frac{3}{4} = \frac{}{100} = \underline{}$ %

There are 100 patches in each of these quilts.

5 (a) Colour 25% red, 75% blue.

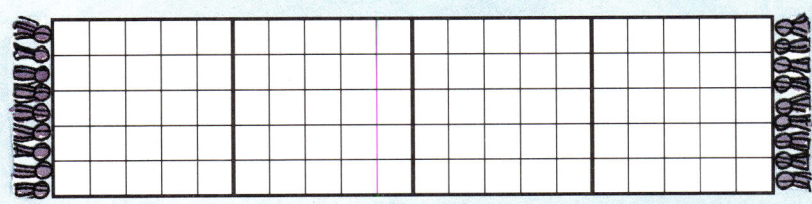

(b) Complete: $25\% = \frac{}{100} = \frac{}{4}$ $75\% = \frac{}{100} = \frac{}{4}$

6 (a) Colour 50% red, 10% blue.

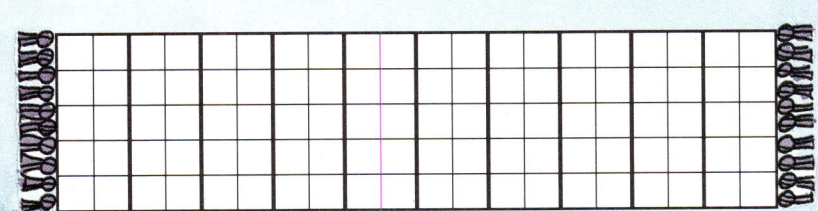

(b) Complete: $50\% = \frac{}{100} = \frac{}{2}$ $10\% = \frac{}{100} = \frac{}{10}$

Go to Textbook page 53.

Function machines

Complete each function machine.

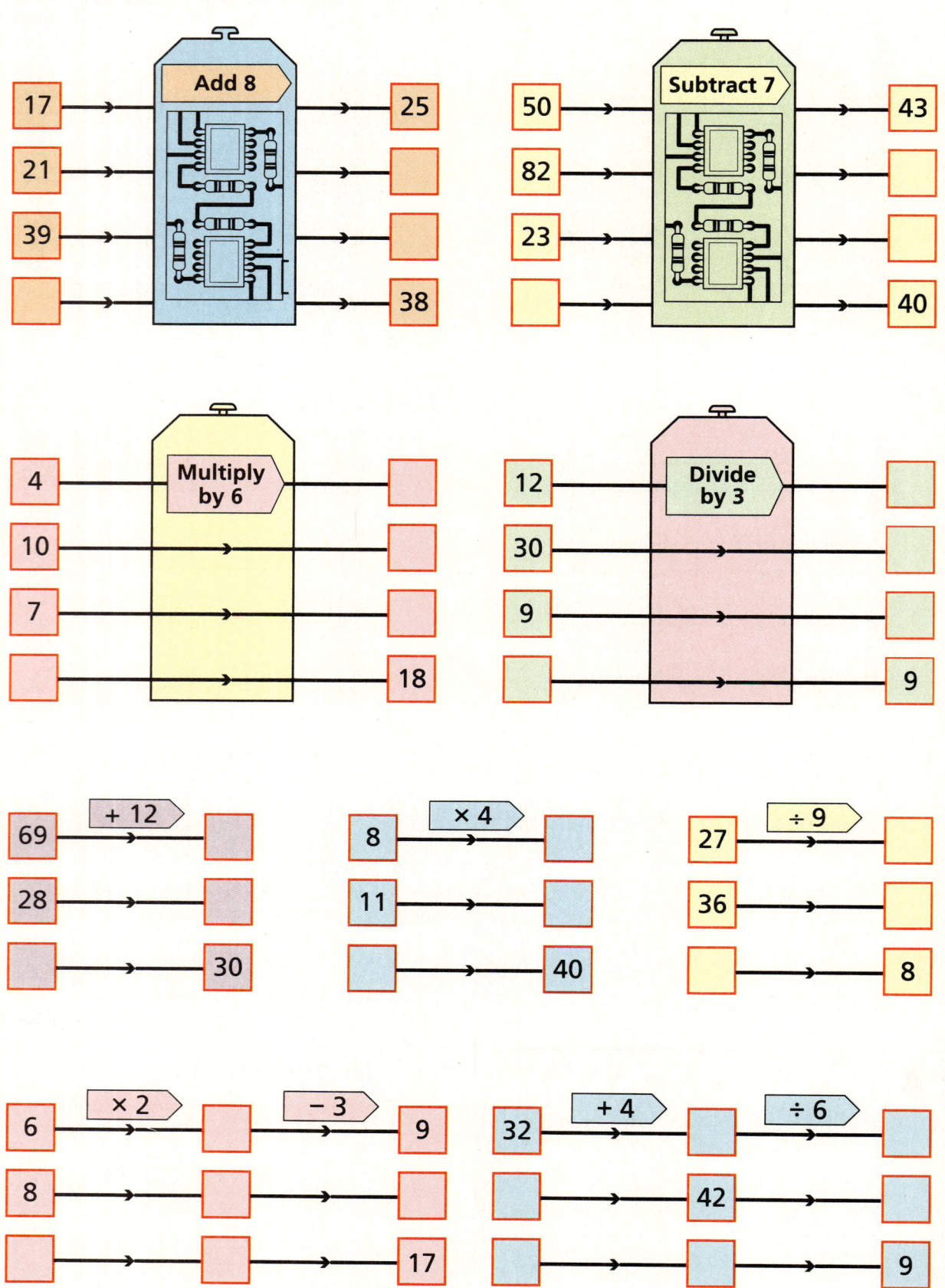

Go back to Textbook page 55.

Matchstick patterns

1 (a) Draw the next two patterns.

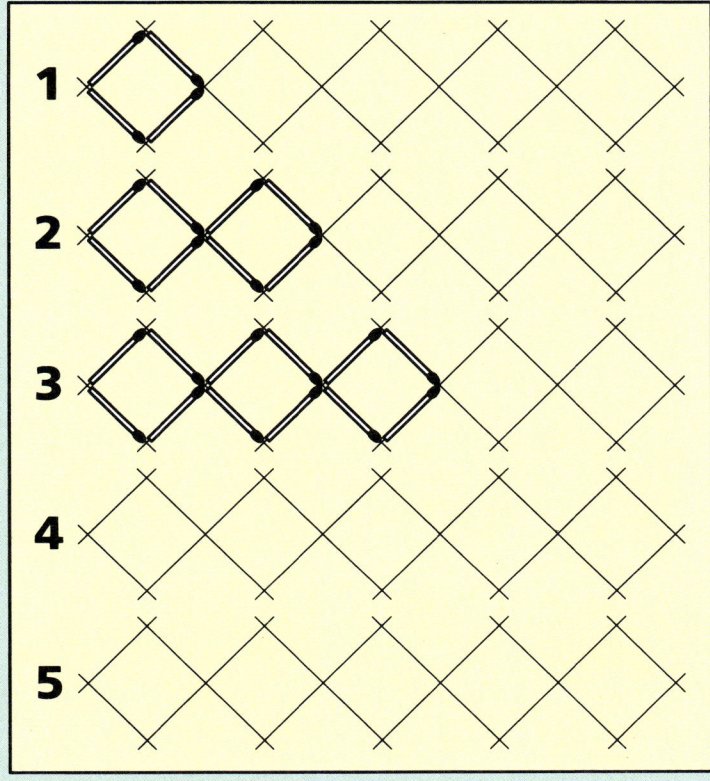

(b) Complete:

Number of squares	Number of matches
1	4
2	
3	
4	
5	

(c) How many matches are needed for 10 squares? _____

(d) Complete:
The number of matches is _____ times the number of squares.

2 (a) Draw the next two patterns.

(b) Complete:

Number of triangles	Number of matches
1	3
2	
3	
4	
5	

(c) How many matches are needed for 20 triangles? _____

(d) Complete:
The number of matches is _____ the number of triangles.

Go to Textbook page 56.

17 Count it out

Length: perimeter, investigation

1. Check the area and perimeter of this shape.

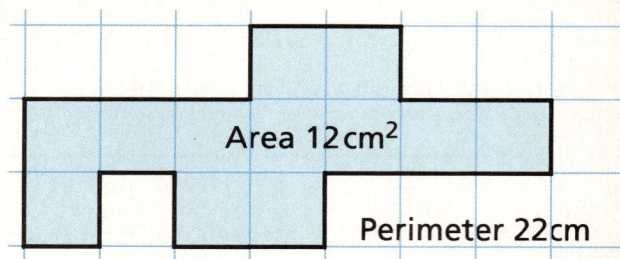

Area 12 cm²

Perimeter 22 cm

2. Draw shapes with an area of 12 cm² and tick (✓) the true statement.

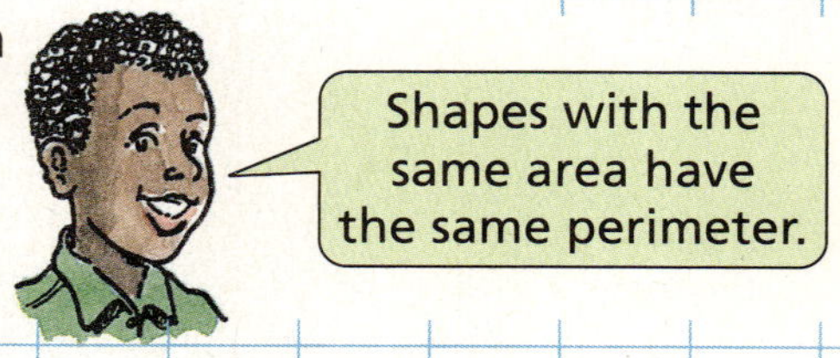

Shapes with the same area have the same perimeter.

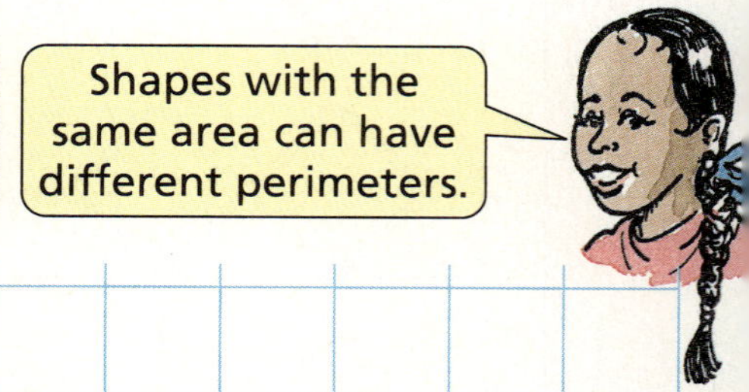

Shapes with the same area can have different perimeters.

3. Draw shapes with an area of 11 cm².
   Try to find the shape with the shortest possible perimeter.

Problem solving

4. Draw 2 different shapes each with
   - an area of 10 cm² **and**
   - a perimeter of 14 cm.

Go to Textbook page 71.

18
2D shape

Tangram for Textbook page 98.

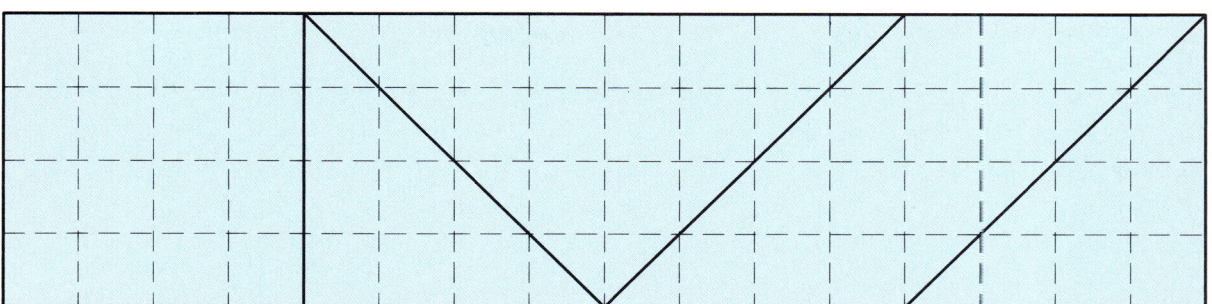

Shapes for Textbook pages 95 and 96.

19

20
2D shape

Triangles for Textbook page 97, question 5.

Triangles for Textbook page 97, questions 3 and 4.

Shapes for Textbook page 96, question 4.

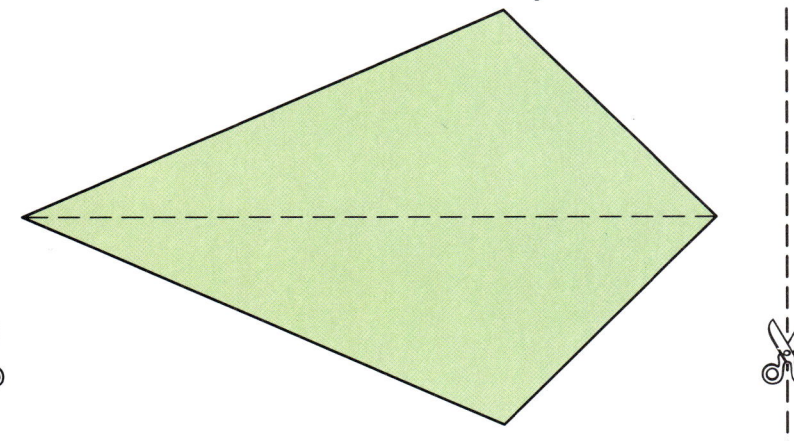
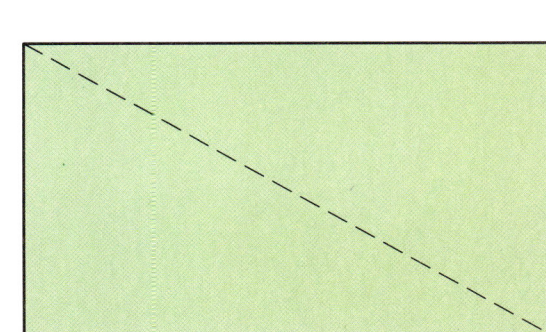

21

Angles for use with Textbook pages 109 and 110.

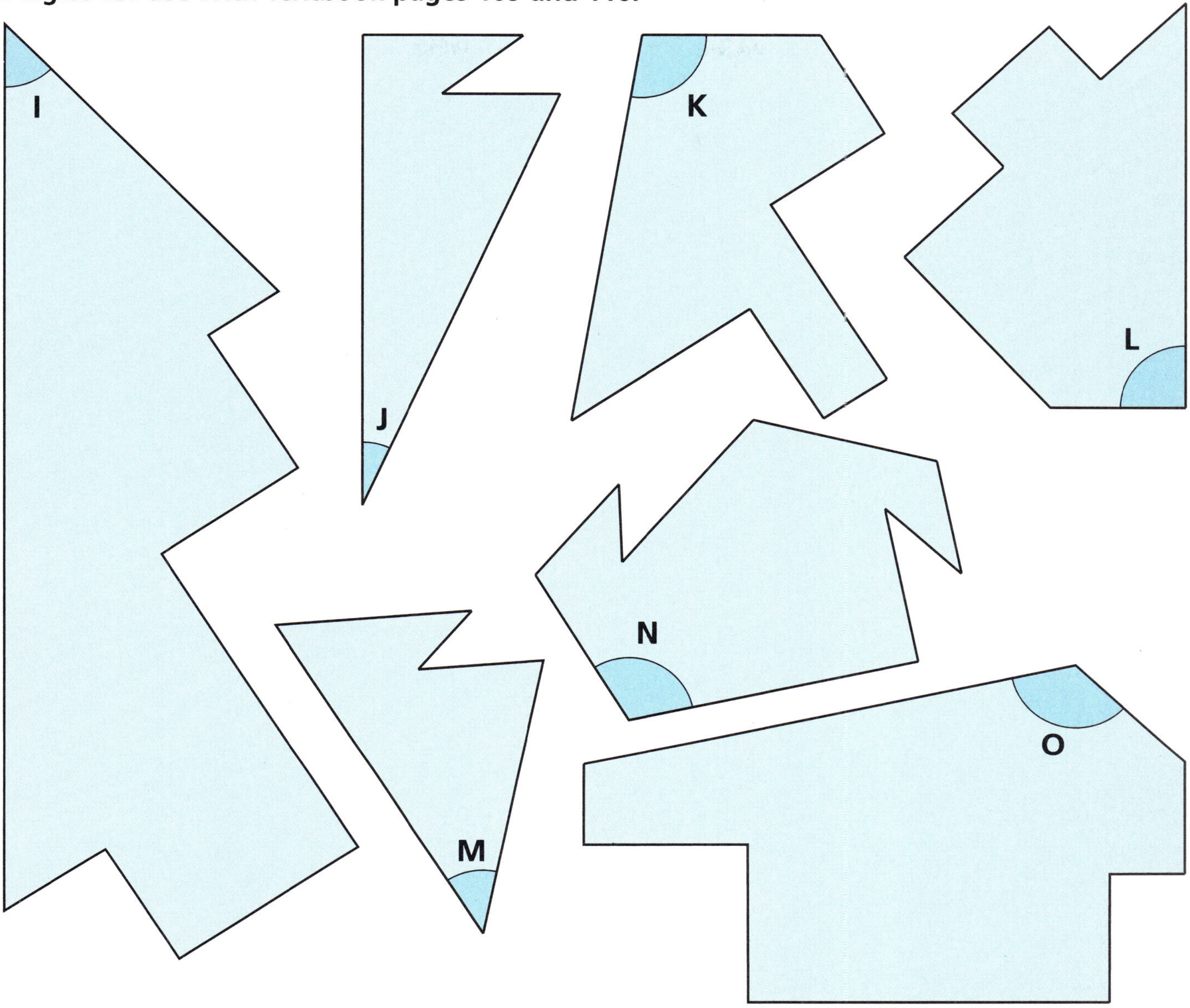

22

Angles: measuring in degrees

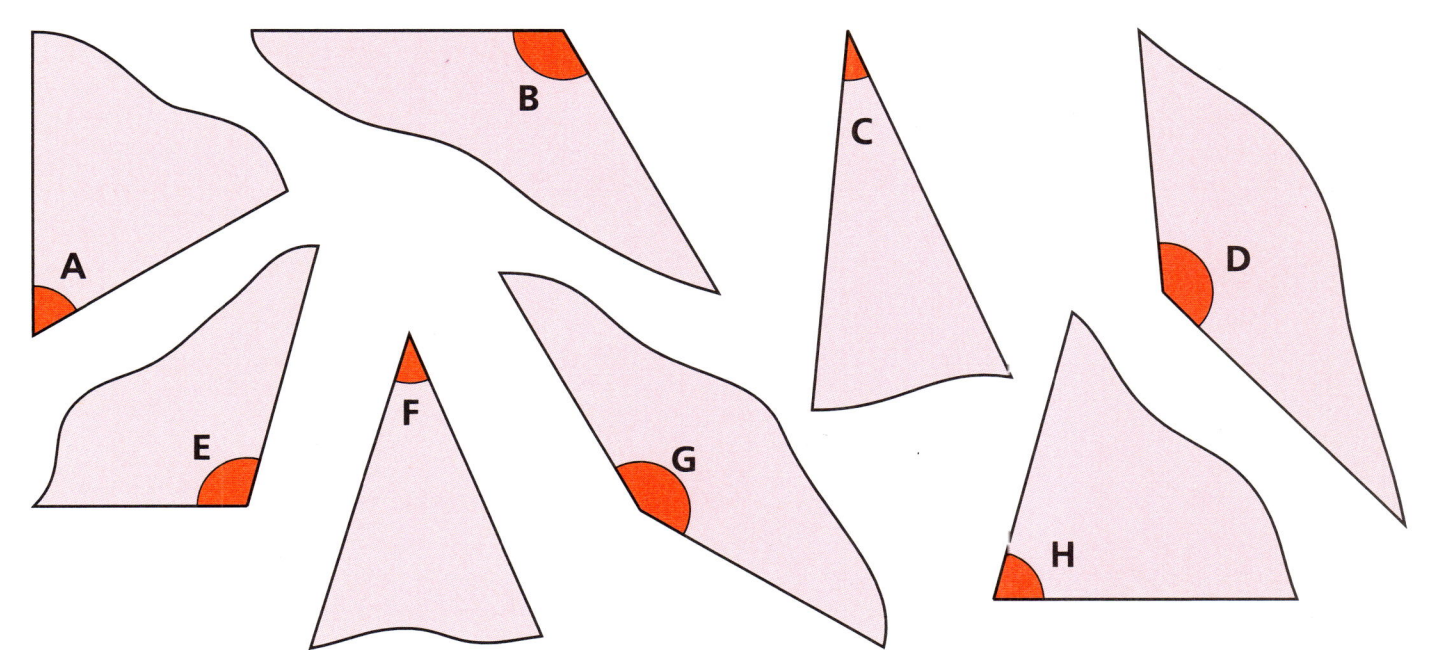

23

Get to the point

24 Weight: scales

Debbie is weighing some ingredients to feed the animals.

1 On the scales, draw pointers to show the weight of each ingredient.

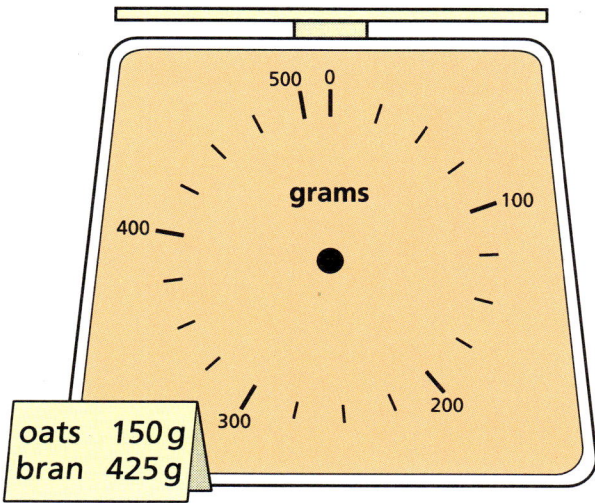

oats 150 g
bran 425 g

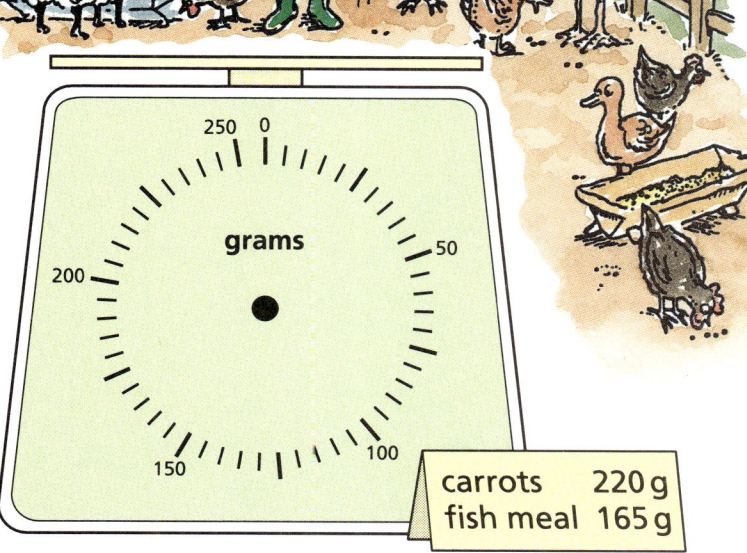

carrots 220 g
fish meal 165 g

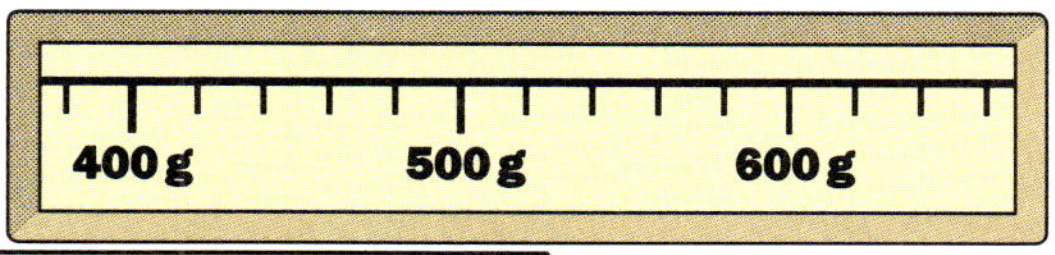

corn 420 g
dried milk 580 g

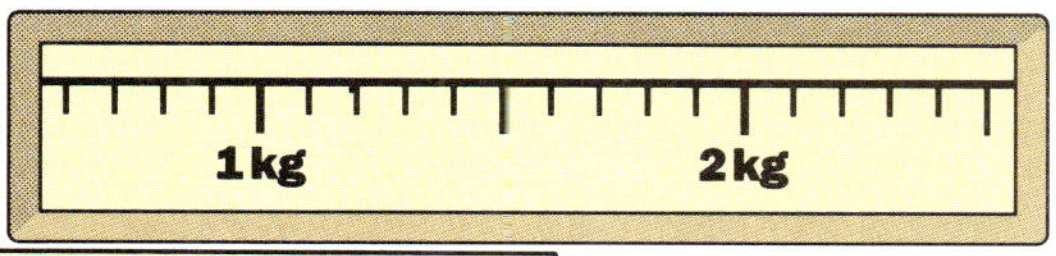

turnip 2 kg 300 g
cabbage 1½ kg

2 These are the weights of puppies at APEC.
For each puppy • choose one of the scales
• draw a pointer to show its weight.

Benny 760 g
Smudge 2 kg 500 g
Patch 3 kg 600 g

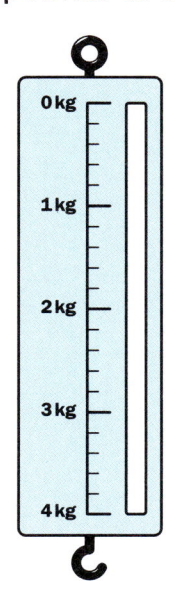
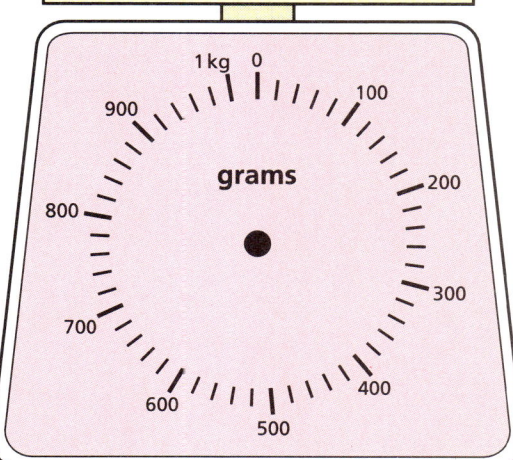

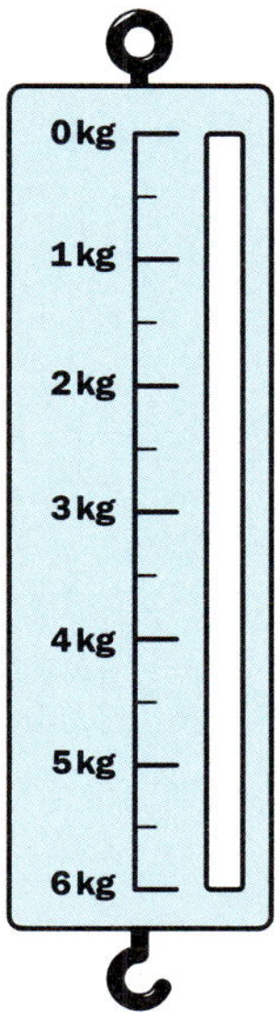

Go to Textbook page 78.

R23 H42

25 Area: irregular shapes

Lands beyond

1 Find the area of each emblem.

Area: _____ cm² Area: _____ cm² Area: about _____ cm²

Area: _____ Area: _____ Area: _____

2 Use centimetre squared paper.
   (a) Design an emblem with an area of about 30 cm².
   (b) Ask a friend to check its area.

26 Area: rectangles

1 For each rectangular door,
- colour the white rows
- complete the table.

Door	Number of rows altogether	Number of squares in each row	Area of door in squares
Jewel Room	6	7	
Alltmouth			
Guard Room			
Way Out			
Gallery			
Cellar			

2 How can you find the area of a rectangle **without counting** squares?

_____

_____

Go to Textbook page 79.

Designs and paintings

1 Divide each design into squares and rectangles and find its area.

Area = _____ squares

Area = _____

Area = _____

Area = _____

2 Find the perimeter and the area of each painting.

Perimeter = _____ cm

Area = _____ cm²

Perimeter = _____

Area = _____

Perimeter = _____

Area = _____

3 What do you notice about

• the perimeters _____

• the areas? _____

Ask your teacher what to do next.

Feeds

1 Complete each scale by writing a volume beside each mark.

(a)

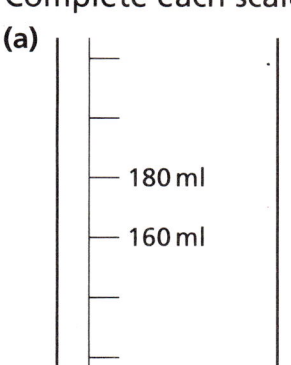

180 ml
160 ml

(b)

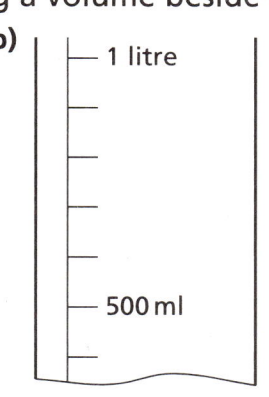

1 litre
500 ml

(c)

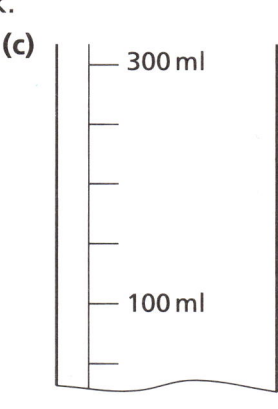

300 ml
100 ml

2 Colour the volumes of these ingredients on the scales in question **1**.

(a) broth: 140 ml **(b)** plant juice: 800 ml **(c)** milk: $\frac{1}{4}$ l

3 Colour the bottles to show the volume in each baby's feed.

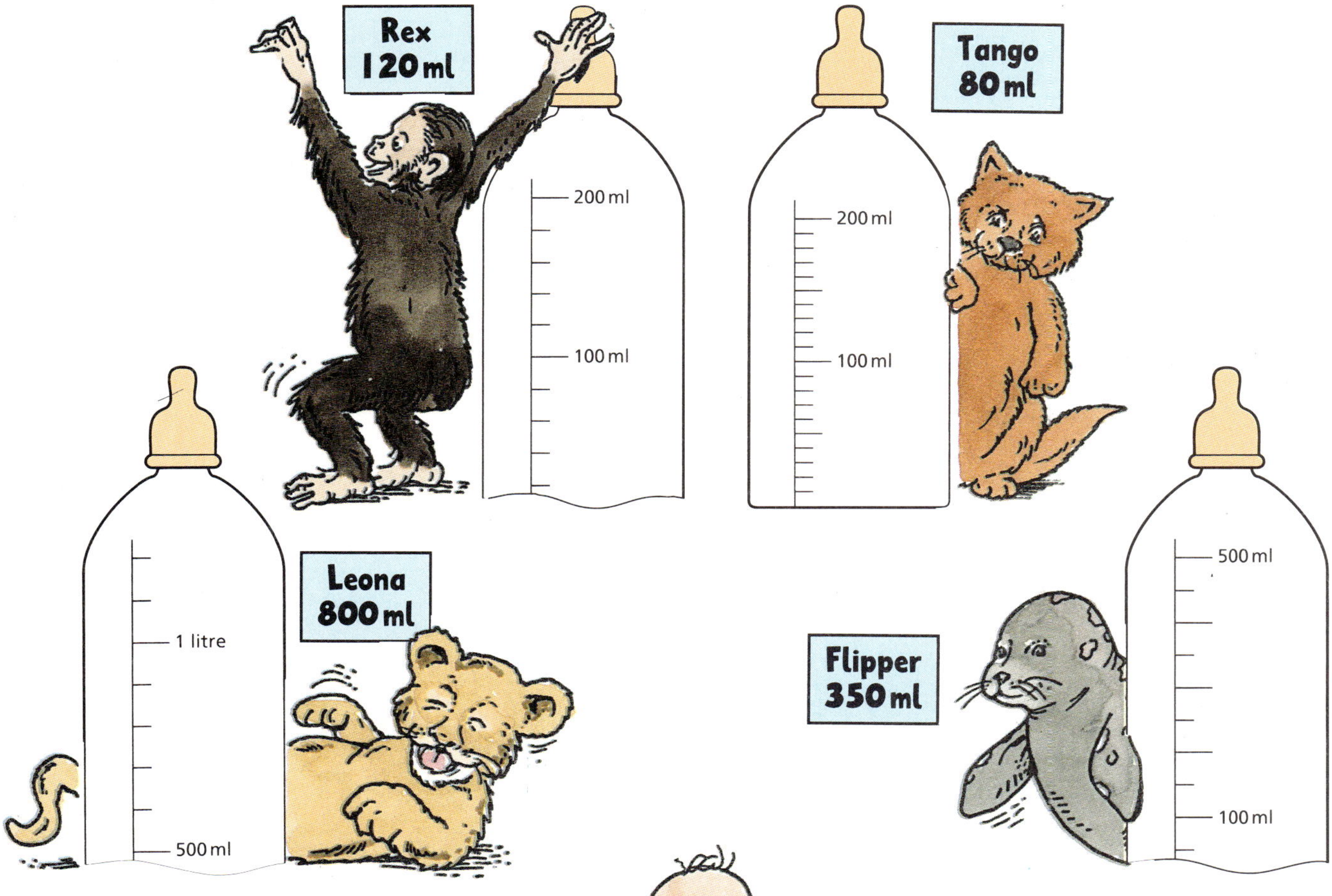

Rex
120 ml
200 ml
100 ml
Tango
80 ml
200 ml
100 ml
1 litre
500 ml
Leona
800 ml
Flipper
350 ml
500 ml
100 ml

4 Use measuring jars and water to measure out feeds for these two babies:

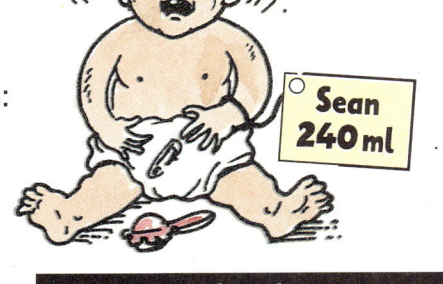

Sean
240 ml

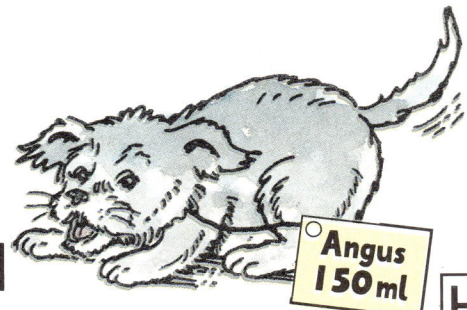

Angus
150 ml

Go to Textbook page 82.

29 Time: 24-hour clock durations

6TV schedules

1 Antonia is making a schedule for these Thursday programmes.

Complete the **afternoon** schedule.

Lunchtime	Film Box
1 hour	2 hours

About Britain	Cartoon
45 minutes	15 minutes

Newsround	Friends
$\frac{1}{2}$ hour	25 minutes

6TV — Thursday afternoon

Start	Programme	Finish
13.00	Lunchtime	
14.00	Film Box	
16.00	About Britain	
16.45		17.00
17.00		17.25
17.25		17.55

Problem solving

2 Complete Antonia's schedule for these **evening** programmes.

Pop Toppers	Showtime
$\frac{1}{2}$ hour	$1\frac{1}{2}$ hours

Snooker	Meet the Mob
1 hour 45 minutes	40 minutes

6TV — Thursday evening

Start	Programme	Finish
	Meet the Mob	
	Pop Toppers	19.30
21.00		22.45

Go to Textbook page 91.

TIMEWARP Challenges

30 — Time: practical timing

In today's programme we want to find out about YOU!

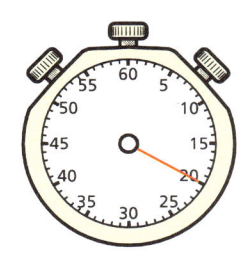

Work with a partner.
Use a watch which measures time in seconds.

1 NIMBLE FINGERS

You have **20 seconds** for each activity.
How many times can you write
all these numbers? _____
0 1 2 3 4 5 6 7 8 9

How many times can you write
Timewarp ? _____

You have **40 seconds** for each activity.
How many tables facts can
you write? _____
6 × 7 = 42   5 × 5 = 25

How many cubes can you
join together? _____

You have **1 minute** for each activity.
How many times can
you throw a 6? _____

How many happy faces can
you draw? _____

2 Who has the **nimblest fingers** in your group?

0 1 2 3 4 5 _____   Timewarp _____   6 × 7 = 42 _____

_____   _____   _____

FANCY FEET

3 How long do you take to
(a) walk heel-to-toe for 10 m   estimate _____   measure _____
(b) bunnyhop 20 times   estimate _____   measure _____
(c) hop across the hall   estimate _____   measure _____
(d) walk twice around the hall?   estimate _____   measure _____

Ask your teacher what to do next.

31 Co-ordinates

Starfighters

1 (a) Make starfighter pictures. Mark these points on the grid and join them, **in order**, with straight lines.

First starfighter (2, 7) (1, 7) (1, 8) (0, 9) (2, 9) (3, 8) (13, 8) (17, 6) (12, 6) (11, 5) (4, 5) (5, 6) (3, 6) and back to (2, 7)

Second starfighter (11, 1) (6, 0) (7, 1) (2, 1) (3, 2) (2, 5) (4, 3) (6, 4) (9, 4) (11, 3) (12, 3) (15, 2) (12, 1) and back to (11, 1)

(b) Colour the starfighters and write a name for each.

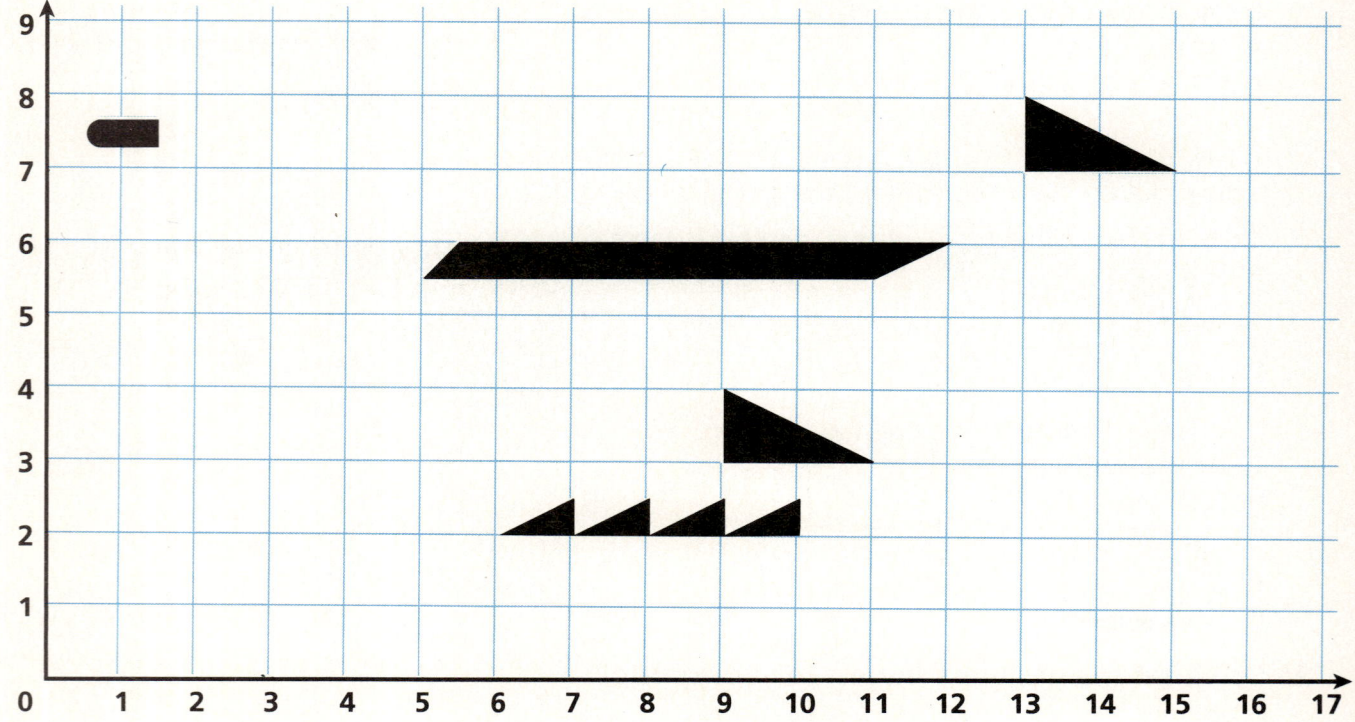

2 Draw your own starfighter on the grid below.
List its co-ordinates, in order, on a piece of paper.
Give this to a friend to draw the starfighter on squared paper.

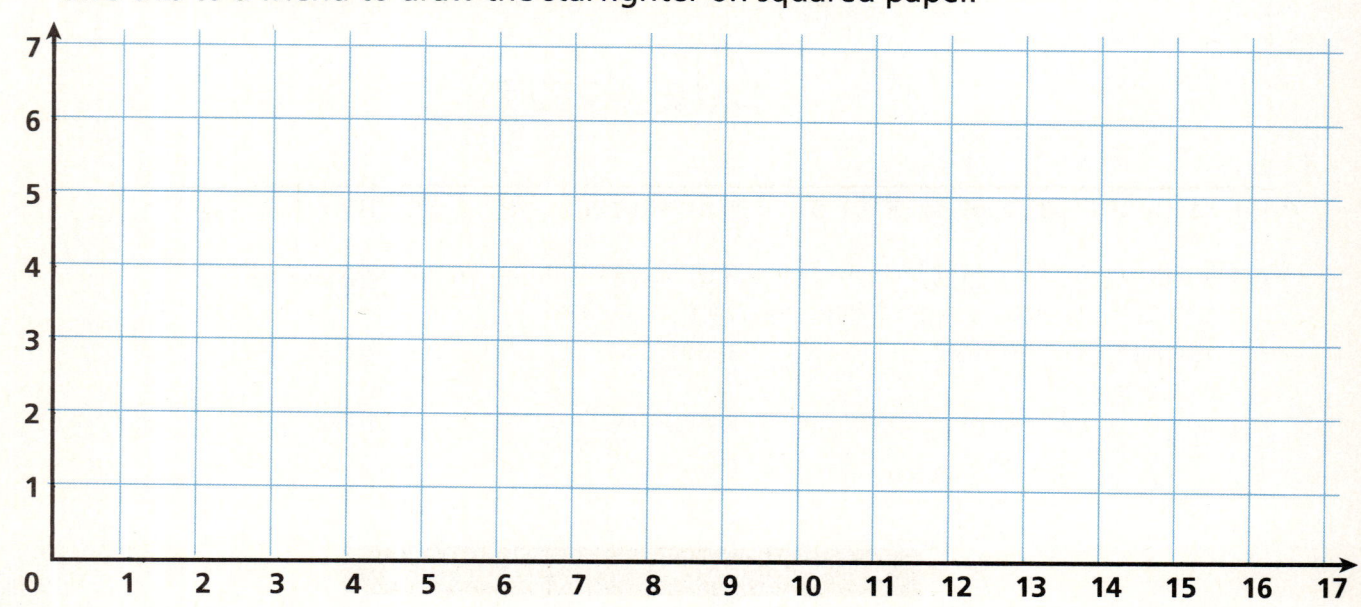

Spacefleet

1 (a) Mark these points on the grid.
Join them, in order, to make a *Spacefleet* badge.

(3, 0) (2, 1) (0, 0) (1, 2) (0, 3) (1, 4) (0, 6)
(2, 5) (3, 6) (4, 5) (6, 6) (5, 4) (6, 3) (5, 2)
(6, 0) (4, 1) and back to (3, 0)

(b) Draw the lines of symmetry on the badge.
Colour it so that it still has these lines of symmetry.

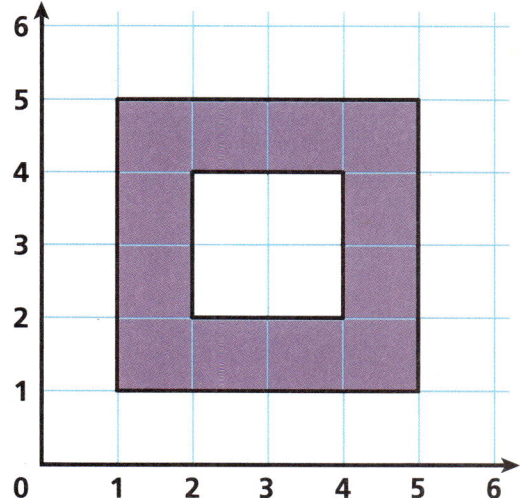

2 (a) Mark each set of points on the grid below and join them, in order, to make **half** of a **symmetrical** picture.

Spacefleet cruiser
(3, 1) (2, 0) (1, 1) (2, 2) (2, 4)
(0, 2) (0, 4) (2, 6) (2, 7) (3, 9)

Spacefleet station
(15, 4) (14, 3) (14, 2) (13, 1) (12, 1)
(11, 0) (10, 1) (9, 1) (8, 2) (8, 3) (7, 4)

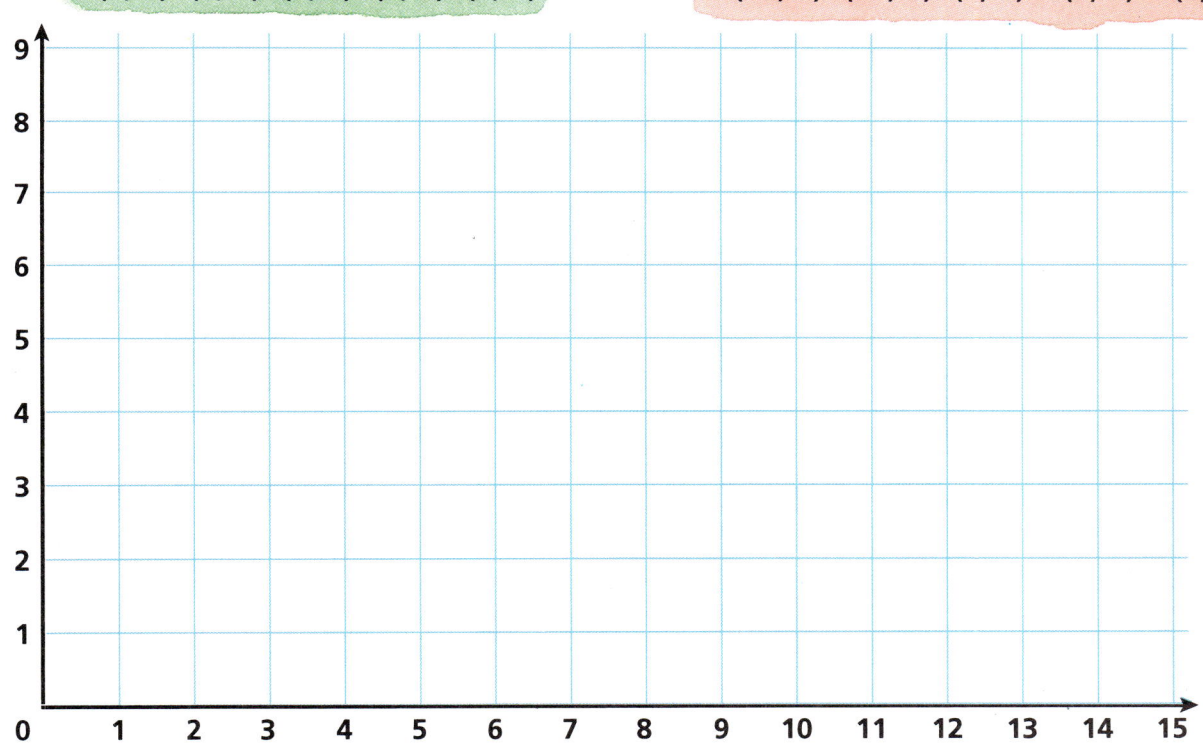

(b) Draw the other half of each picture by marking and joining points.

(c) List the co-ordinates of these points in order.

cruiser:
(3, 9) _____

station:
(7, 4) _____

Ask your teacher what to do next.

33 — 2D shape: drawing shapes

Figure it out

1 For each set of co-ordinates
   • mark the points
   • join them, in order, to make a shape
   • name the shape.

   (a) (1, 0)   (1, 4)   (3, 5)   (3, 1)
   (b) (4, 2)   (7, 5)   (9, 3)   (6, 0)
   (c) (9, 1)   (11, 5) (13, 3)
   (d) (14, 1) (15, 4) (18, 5) (17, 2)

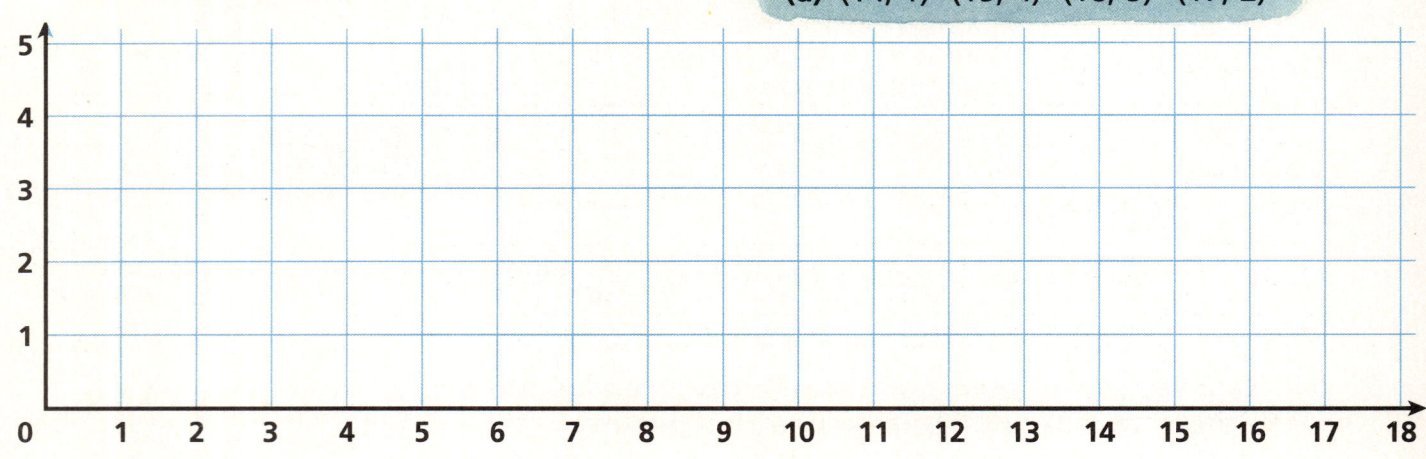

Problem solving

2 (a) On each grid mark these 3 points: (2, 1) (1, 4) (2, 5).
   (b) Make the given shape by marking one more point and joining up the four points.
   (c) Draw the diagonals of each shape.

kite

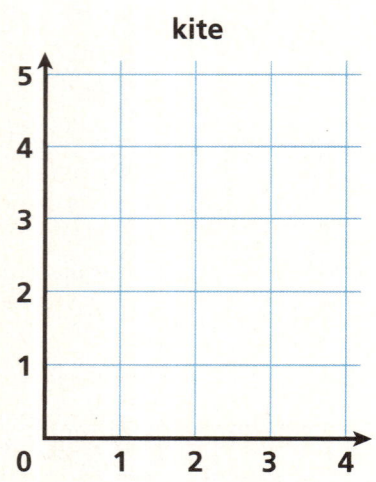

parallelogram

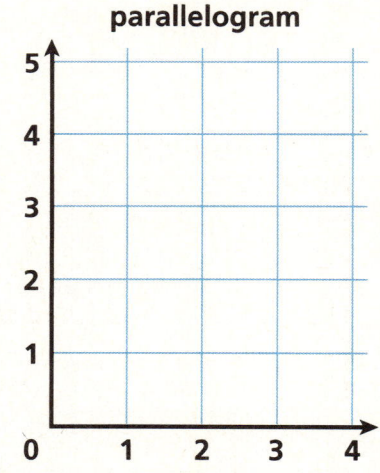

four-sided shape with 2 right angles

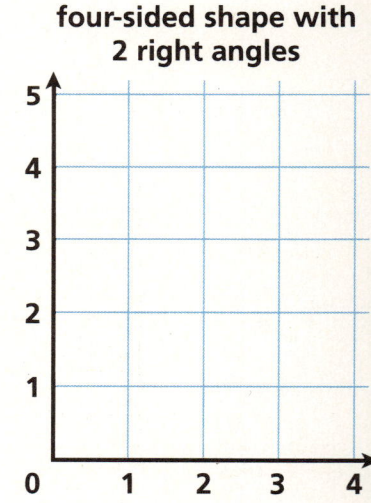

Problem solving

3 (a) Draw squares like this on centimetre squared paper.
   (b) In each square draw a different four-sided shape. The four corners of each shape must be on different sides of the square.
   (c) Name each shape you make.

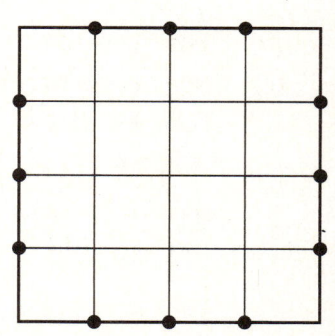

Go to Textbook page 98.

Top class mural

34 Handling data: pictograms, bar graphs

The children paint a mural on the playground wall. They keep a record of the number of bricks painted.

Each ▭ represents 10 bricks.

Each ▫ represents 5 bricks.

1 Complete the table.

Colour	Number of bricks painted	Total
red	10 10 10 10 10 10 10 10 10	
blue	10 10 10 10 10 10 10 10 10 5	
green	10 5 10 10 10 10 10	
purple	10 10 10 5	
yellow	10 10 10 5 10 10 10 10 10	

2 Which colour was used to paint

  • exactly one fifth of the bricks _____

  • less than one tenth of the bricks? _____

3 Complete this bar graph using the information in the table.

Colours of painted bricks

Number of bricks

50

0

red   blue   green   purple   yellow

Go to Textbook page 112.

35 A green sweep

Handling data: class intervals

These are the weights of the bags of rubbish collected by Class 6.

4 kg	7 kg	10 kg	19 kg	13 kg	12 kg
16 kg	14 kg	11 kg	9 kg	17 kg	14 kg
8 kg	11 kg	20 kg	5 kg	13 kg	16 kg
3 kg	12 kg	12 kg	22 kg	15 kg	13 kg
8 kg	10 kg	18 kg	15 kg	11 kg	6 kg

1 (a) Use the weights of rubbish to complete the yellow table.
   (b) Complete the yellow bar graph.

2 (a) Use the weights again to complete the green table on **Workbook page 36**.
   (b) Complete the green bar graph.

3 Do this again for the blue table and bar graph on **Workbook page 36**.

4 For each graph give the number of class intervals.

   yellow _____

   green _____   blue _____

5 Discuss the graphs with your teacher.

Weight in kg	Tally marks	Total
1–6		
7–12		
13–18		
19–24		

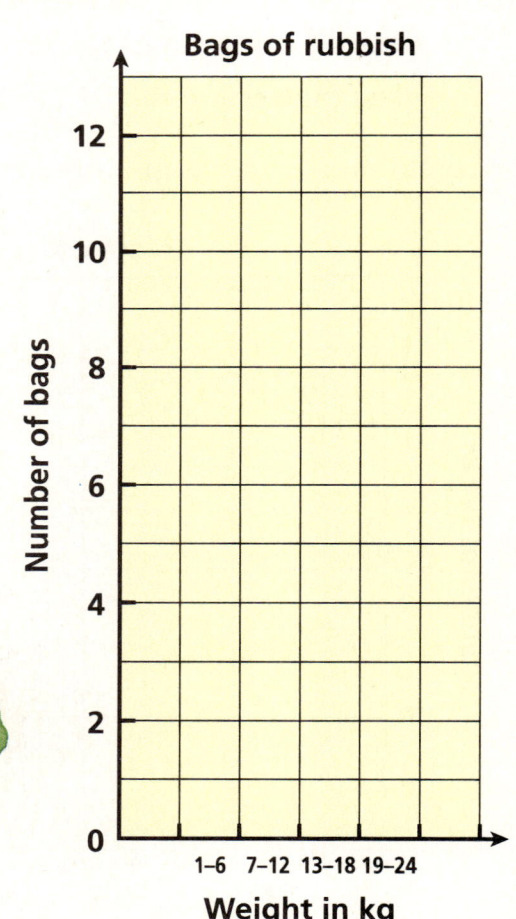

Go to Textbook page 118, question 2.

36
Handling data: class intervals

Weight in kg	Tally marks	Total
1–4		
5–8		
9–12		
13–16		
17–20		
21–24		

Weight in kg	Tally marks	Total
1–3		
4–6		
7–9		
10–12		
13–15		
16–18		
19–21		
22–24		

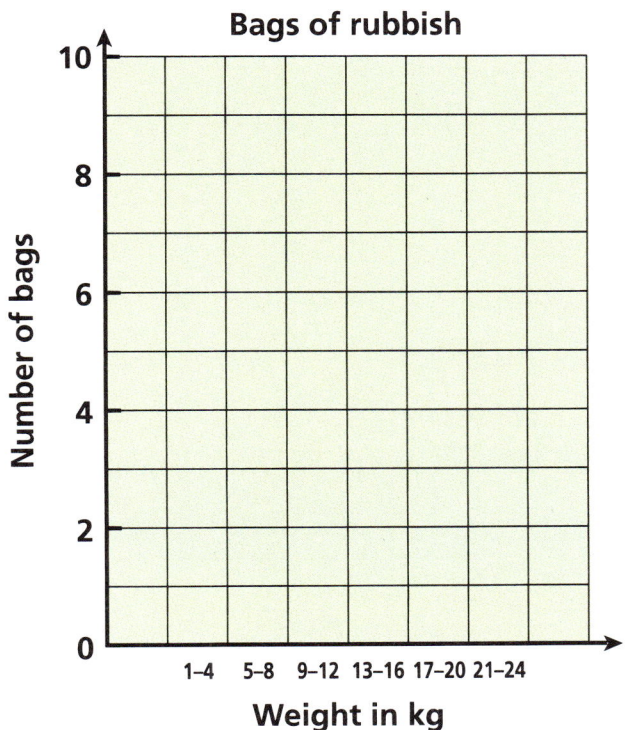

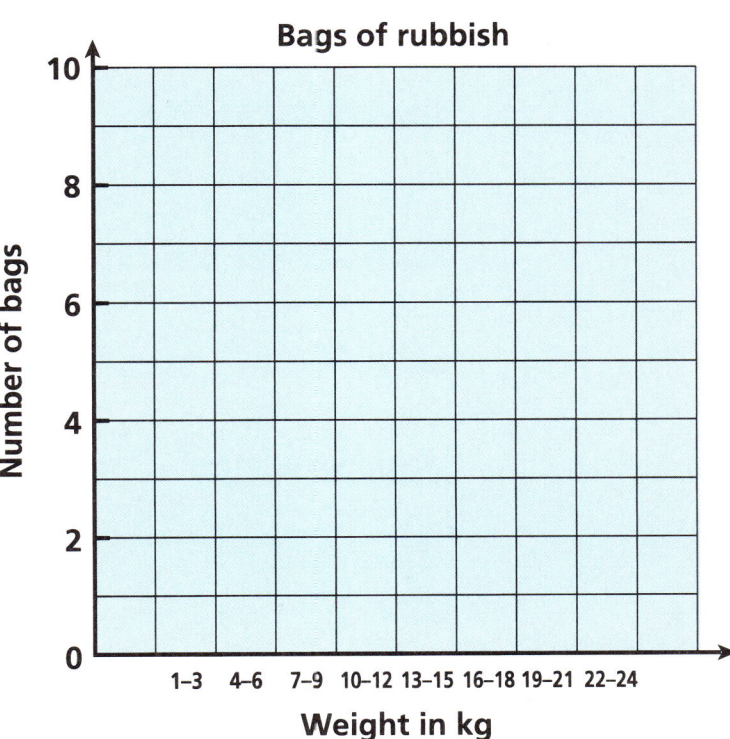

Number: Record of Work

HEINEMANN MATHEMATICS 6

Name _____ Class _____

Textbook / Workbook / Reinforcement Sheets / Check-ups   Extension Textbook

Place value, addition and subtraction
| T1 | T2 | W1 | R1 | | W2 | W3 | T3 | R2 | | Check-up 1 |

| T4 | R3 | T5 | Check-up 2 | | | | | E2 | E3 |

Rounding and estimation
| T6 | T7 | R4 | T8 | Check-up 3 | | | | E4 |

Calculator
| T9 | T10 |

Multiplication
| T12 | W4 | W5 | T13 | | | | E6 |

Division
| T14 | T15 | T16 | R5 | | T17 | T18 | Check-up 4 | | E8 |

| T19 | T20 | T21 | R6 |

| T22 | T23 | T24 | R7 | T25 | W6 | Check-up 5 | | E7 |

| T26 | W7 | R8 | T27 | T28 | R9 | Check-up 6 |

Fractions
| T30 | W8 | T31 | R10 | W9 | T32 | T33 | R11 |

| T34 | T35 | Check-up 7 | | | | | E9 |

Decimals
| T36 | T37 | T38 | T39 | T40 | T41 | Check-up 8 |

| T42 | W10 | R12 | T43 | T44 | R13 | W11 | Check-up 9 |

| T45 | T46 | T47 | R14 | T48 | T49 | T50 | T51 | Check-up 10 | Check-up 11 | E10 | E11 |

Percentages
| T52 | W12 | R15 | W13 | W14 | T53 | R16 | Check-up 12 | | E13 |

Pattern
| W15 | T55 | W16 | R17 | T56 | T57 | T58 | | E14 | E15 |

Number

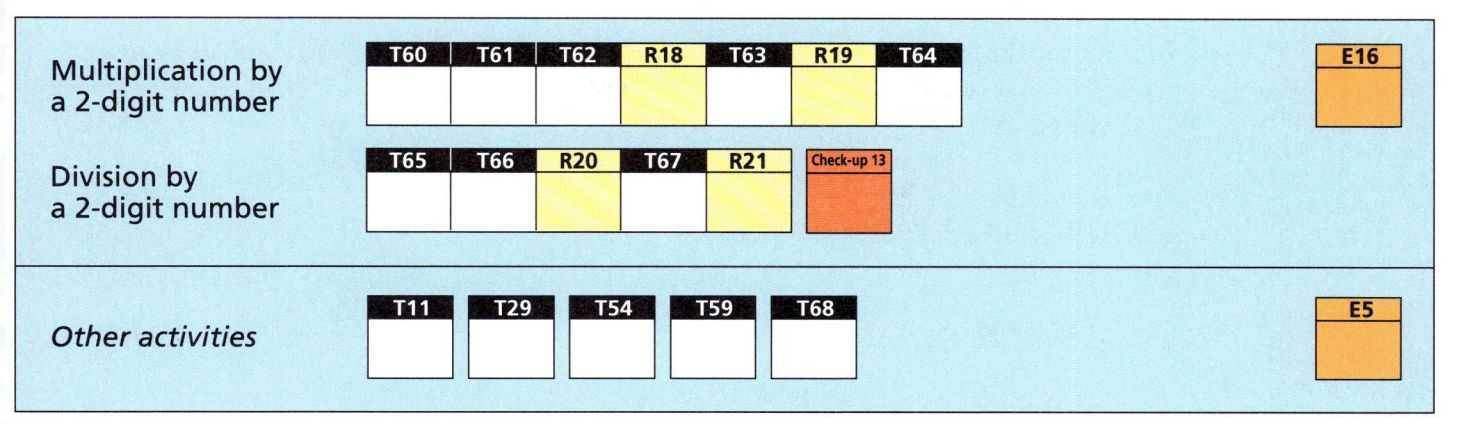

Textbook / Workbook / Reinforcement Sheets / Check-ups   **Extension Textbook**

Multiplication by a 2-digit number	T60	T61	T62	R18	T63	R19	T64		E16
Division by a 2-digit number	T65	T66	R20	T67	R21	Check-up 13		

Other activities	T11	T29	T54	T59	T68		E5

Measure: Record of Work

Name _____ Class _____

Textbook / Workbook / Reinforcement Sheets / Check-ups   **Extension Textbook**

Length	T69	T70	R22	W17	T71	T72	T73	T74	Check-up 1	E17
Weight	T75	T76	T77	W24	R23	T78	Check-up 2		

Area	W25	W26	T79	T80	W27	R24	Check-up 3	E18

Volume	T81	W28	T82	R25	T83	T84	Check-up 4	E19

Measure	E20 E21

Time	T85	T86	R26	T87	T88	R27	Check-up 5	
	T89	T90	R28	W29	T91	W30	Check-up 6	E22 E23

Other activities	T92	T93

Shape: Record of Work

HEINEMANN MATHEMATICS 6

Name _____ Class _____

Textbook / Workbook / Reinforcement Sheets / Check-ups　　　Extension Textbook

Co-ordinates
| T94 | W31 | W32 | Check-up 1 | | | | | E24 |

2D shape
| T95 | W18 | T96 | W20 | T97 | W33 | R29 | T98 | Check-up 2 | E26 |

| T99 | T100 | T101 | T102 |

3D shape
| T104 | T105 | T106 |

Angles
| T108 | T109 | W22 | T110 | R30 | Check-up 3 |

Other activities
| T103 | T107 | | | | E1 | E12 | E25 | E27 | E30 |

Handling data: Record of Work

Name _____ Class _____

Textbook / Workbook / Reinforcement Sheets / Check-ups　　　Extension Textbook

Handling data
| T111 | W34 | T112 | T113 | T114 | T115 | T116 | T117 | Check-up 1 |

| T118 | W35 | W36 | T119 | T120 | T121 | T122 | Check-up 2 | E28 | E29 |

Probability
| T123 | T124 | T125 | Check-up 3 |

Assessment　　Round-up 1　　　Round-up 2　　　Round-up 3

Published by Heinemann Educational Publishers, Halley Court, Jordan Hill, Oxford OX2 8EJ, a division of Reed Educational and Professional Publishers Ltd.
Single ISBN 0 435 02225 3 © Scottish Primary Mathematics Group 1995.
First published 1995.　　03 11
Designed by Miller, Craig & Cocking. Produced by Oxprint Ltd, Oxford.
Illustrated by Oxford Illustrators. Printed by Pindar plc, Scarborough.